川哥教你 Spring Boot 2 实战

◎ 李世川 编著

清华大学出版社
北 京

内 容 简 介

本书重点关注和介绍 Spring Boot 2 框架的技术应用。Spring Boot 是目前微服务架构使用最为广泛的框架之一，一旦开发人员掌握了该框架的配置和使用技巧，则对于当前微服务应用的开发将得心应手。本书详细介绍了 Sping Boot 2 的开发环境、开发方式、数据库应用、MVC、RESTful、安全及测试等。对于刚开始接触微服务架构、Spring Boot 的开发人员来说，这是一本很适合的书籍。本书介绍的内容较多，但都是当前基于微服务架构的应用所需，内容翔实并提供所有源码。对于已有开发经验的 Spring 开发人员来说，这也是一本有价值的参考书。本书案例源码均由作者亲自编写，其中包含了很多有用的方法与使用技巧。

本书适合初级、中级 Java 开发工程师，以及从其他开发语言（如 PHP、C、Python 等）转向 Java 微服务的开发工程师使用，也可作为互联网应用研发人员、自学 Java 开发的大中专院校学生的参考读物。

本书封面贴有清华大学出版社防伪标签，无标签者不得销售。
版权所有，侵权必究。举报：010-62782989，beiqinquan@tup.tsinghua.edu.cn。

图书在版编目(CIP)数据

川哥教你 Spring Boot 2 实战/李世川编著.—北京：清华大学出版社，2023.6
ISBN 978-7-302-62567-4

Ⅰ.①川… Ⅱ.①李… Ⅲ.①JAVA 语言－程序设计 Ⅳ.①TP312.8

中国国家版本馆 CIP 数据核字(2023)第 022873 号

责任编辑：贾　斌
封面设计：何凤霞
责任校对：焦丽丽
责任印制：丛怀宇

出版发行：清华大学出版社
网　　址：http://www.tup.com.cn，http://www.wqbook.com
地　　址：北京清华大学学研大厦 A 座　　　邮　编：100084
社 总 机：010-83470000　　　邮　购：010-62786544
投稿与读者服务：010-62776969，c-service@tup.tsinghua.edu.cn
质量反馈：010-62772015，zhiliang@tup.tsinghua.edu.cn
课件下载：http://www.tup.com.cn，010-83470236

印 装 者：三河市少明印务有限公司
经　　销：全国新华书店
开　　本：185mm×260mm　　印　张：18　　字　数：435 千字
版　　次：2023 年 6 月第 1 版　　印　次：2023 年 6 月第 1 次印刷
印　　数：1～2500
定　　价：79.80 元

产品编号：089779-01

前言

　　距离作者上一本书的发行，又过去了两年。在这两年中，计算机技术、互联网技术高速发展，尤其计算机开发语言发展之快，让人应接不暇。虽然开发人员选择余地变大，但同时陷于各种技术的比较，而无法沉下心来提高技术。作者从事计算机行业多年，热衷于用各种编程语言开发应用系统、App等，但Java编程语言一直具有很大的吸引力，让人一旦用上，就无法放下。

　　查看每年计算机编程语言在全世界的应用排行，Java始终是最流行的语言之一，很欣慰，我一直在使用，当你在看本书时，恭喜你也加入了最大的编程语言阵营。世界上很多大公司都在使用Java，我的周围有80％的人和公司也在使用Java开发应用系统、框架和平台等。Java如此受欢迎，主要归功于其可移植性强、可扩展性灵活、可用插件和开源框架丰富，以及用户社区庞大。

　　作者在编写本书之前，一直用Spring Boot 2开发应用系统，惊讶于其开发如此便捷，可能只需简单单击鼠标，便可以使其运行起来。Spring Boot 2是当前Java开发中比较流行的框架，其是Spring框架的扩展，目标是快速、高效开发基于Java的应用系统，方式是约定大于配置。

　　本书和作者上一本书最大的区别便是本书使用了当前最新的技术，应用Maven技术简化Jar包管理，使得Spring Boot 2的开发变得更加简单。基于此，本书能提供更多示例。

　　当今，Java开发已变得令人烦躁，开发人员不想再局限于传统式开发，即编写一大堆配置文件，手动管理Jar包依赖，以及在开发中不停重新启动应用进行代码调试。在没有使用Jar包管理工具前，找到Jar包依赖是一件很痛苦的事，特别是想升级依赖Jar包时，更是如此；在开发之初，为了启动应用，需要先进行各种配置，可能会产生各种错误，解决这些问题也是很痛苦的事。现在提倡敏捷开发，所见即所得，让开发变成一件快乐的事。Spring Boot 2便是这样一个框架，不失其优雅性、敏捷性，开发人员利用它可以快速开发一个应用、模块或功能，并实现快速部署。

　　本书立足Spring Boot 2框架的实战操作，以作者多年的经验告诉大家，实际操作是掌握编程语言、框架的最佳方法。至少作者本人认为如此。本书通俗易懂，尽量抛开复杂而不易懂的理论，立足实战。本书已涉及Spring Boot 2中多方面，知识点很多，并提供详细案例，目的在于使读者通过本书的学习，快速掌握这个框架，应用于实际项目。

本书共包含 13 章，各章主要内容如下：

第 1 章是本书的基础，介绍 Spring、Spring Boot 2、示例对比，以及开发 Spring Boot 2 的工具。目的在于使读者快速对 Spring Boot 2 形成一个基本概念，并提高对开发的兴趣。

第 2 章介绍 Spring Boot 2 相关的重要基础知识，主要是一个完整的生命周期，即创建、启动、配置、运行、打包等，使读者从整体上掌握 Spring Boot 2 的开发流程。

第 3 章介绍 Spring Boot 2 中的页面处理技术，其不局限于某一种具体页面的处理方式和方法，而是提供了一种灵活的方式，允许开发人员根据需要进行选择。本章也是如此，不聚焦于某一种具体技术，而是介绍了几种当前主流的页面处理技术。在实际开发中，开发人员可以根据需要进行灵活选择。

第 4 章介绍数据操作中的 Spring JDBC 方式，数据操作是应用系统中的重要部分。本章有两个知识点。第一个知识点是内存数据库 H2，这是贯穿本书的重要数据库。其重要体现在两方面：一是可以在不做任何配置和安装的情况下，启动内存数据库 H2 进行代码开发；二是 H2 数据库在测试中占有很重要的地位。所以，本书介绍一些知识点时，都使用数据库 H2 进行讲解。本章第二个知识点是 Spring JDBC 操作数据库，这是 Spring 提供的一种重要的操作关系数据库的方式。

第 5 章介绍数据操作中的 Spring Data JPA 技术。Spring Data 意在统一访问数据接口，Spring Data JPA 加快访问数据的开发，本章详细介绍这种技术，从简单操作到复杂自定义操作多种方式。

第 6 章介绍数据操作中的 MyBatis 技术。MyBatis 数据访问框架更接近于原生 SQL 访问关系数据库，更加适合喜欢 SQL 的开发人员，本章介绍 MyBatis 基本使用方法、高级使用技术以及其支持的注解方式。

第 7 章介绍 Spring Boot 2 中连接多种关系数据库的方法。结合前面第 4~6 章介绍的关系数据操作方法，开发人员可以轻松操作多种数据库。

第 8 章介绍 Spring Boot 2 中连接和操作非关系数据库 MongoDB 的技术。MongoDB 是目前应用较广的非关系数据库，特别适合当前互联网上大数据的应用。通过本章的学习，开发人员可以掌握 Spring Boot 2 中操作 MongoDB 的方法。

第 9 章介绍 Spring Boot 2 中的 MVC 模式。利用当前流行的 MVC 开发模式，可以快速开发出 Spring MVC 应用。本章知识比较综合，须结合前面章节介绍的页面处理技术和数据库操作技术。

第 10 章介绍 Spring Boot 2 中的 RESTful 操作。目前互联网流行前后端分离开发，便于团队中的开发人员发挥各自所长。本章介绍了 RESTful 开发方法和建议的 RESTful 开发规范。

第 11 章介绍 Spring Boot 2 中的安全认证。简单理解就是任何一个互联网应用都需要用户登录后才能操作，Spring Boot 2 中提供了安全认证方式和方法。本章详细介绍该技术和应用，便于开发人员进行选择。

第 12 章介绍 Spring Boot 2 中的测试方法，测试是开发中的重要一环。Spring Boot 2 可以很好地结合当前主流测试框架 JUnit 进行测试工作。本章详细介绍 Spring Boot 2 中结合 JUnit 5 的测试方法。

第 13 章介绍 Spring Boot 2 中其他几个重要的方法，包括上传文件、Lombok 应用、Devtools 应用、更改应用启动标志和缓存的应用。这几个方法在实际开发中都很有帮助，本章将其整合在一起进行介绍。

本书内容涉及 Spring Boot 2 中很多重要的方面，也是开发人员在实际开发中经常遇到的。因本书内容侧重于实践，且对开发具有参考价值，故本书取名"实战"。

在本书的编写过程中，Spring Boot 2 版本更新很快，在成稿之前，作者已将本书所有源码中 Spring Boot 2 版本统一更新至 2.2.6，并进行了相关调试，书中相关内容同步进行了替换。

本书得以快速完成，要感谢家人在我写本书时给予的无私支持，同时感谢同事提供的有力帮助。

作 者

2023 年 5 月

目录

第 1 章 初识 Spring Boot 2 ········· 1
- 1.1 微服务介绍 ········· 1
- 1.2 Spring 介绍 ········· 3
- 1.3 Spring 简单示例 ········· 5
- 1.4 Spring Boot 介绍 ········· 11
- 1.5 Spring Boot 2 示例 ········· 12
- 1.6 Maven POM 文件介绍 ········· 15
- 1.7 使用 Eclipse ········· 18
- 1.8 使用 Spring Tool Suite 4 ········· 23
- 本章小结 ········· 28

第 2 章 启动 Spring Boot 2 ········· 29
- 2.1 启动类 MainApplication ········· 29
- 2.2 Spring Boot Starters ········· 30
- 2.3 相关依赖 ········· 31
- 2.4 配置文件 ········· 32
- 2.5 @RestController 和 @Value ········· 33
- 2.6 logging ········· 35
- 2.7 运行工程 ········· 38
- 2.8 打包工程 ········· 39
- 本章小结 ········· 41

第 3 章 页面处理 ········· 42
- 3.1 Thymeleaf 介绍 ········· 42
- 3.2 集成 Thymeleaf ········· 43

3.3 Thymeleaf 语法 ··· 45
   3.3.1 表达式语法 ······································ 45
   3.3.2 判断 ··· 46
   3.3.3 循环 ··· 47
   3.3.4 属性修饰符 ······································ 48
   3.3.5 内嵌对象 ·· 49
   3.3.6 基本配置 ·· 49
3.4 Thymeleaf 示例 ··· 50
3.5 体验 FreeMarker ······································· 53
3.6 FreeMarker 语法 ······································· 56
   3.6.1 基本规则 ·· 56
   3.6.2 字符输出 ·· 57
   3.6.3 数字格式输出 ···································· 57
   3.6.4 日期格式输出 ···································· 58
   3.6.5 其他数值 ·· 58
   3.6.6 运算符 ··· 59
   3.6.7 页面变量 ·· 59
   3.6.8 判断指令 ·· 60
   3.6.9 循环遍历 ·· 61
   3.6.10 基本设置 ······································· 62
3.7 FreeMarker 示例 ······································· 63
3.8 JSP 介绍 ·· 68
3.9 JSP 语法 ·· 68
   3.9.1 核心标签 ·· 70
   3.9.2 格式化标签 ······································ 70
   3.9.3 JSTL 函数 ······································· 71
   3.9.4 Spring 标签库 ··································· 71
3.10 JSP 示例 ··· 73
本章小结 ··· 84

第 4 章 数据操作——使用 Spring JDBC ············ 85

4.1 H2 数据库 ·· 85
4.2 Java 连接 H2 数据库 ·································· 87
4.3 Spring Boot 2 中 JDBC 连接方式 ··················· 87
4.4 Spring JDBCTemplate ································ 91
本章小结 ··· 96

第 5 章 数据操作——Spring Data JPA ············· 97

5.1 JPA 介绍 ·· 97

5.2　Spring Data JPA ·· 97
5.3　JpaRepository＜T，ID＞方法 ································· 101
5.4　接口规范名方法 ·· 103
5.5　@Query ··· 109
5.6　多表查询 ··· 110
　　5.6.1　一对多映射 ··· 110
　　5.6.2　一对一映射 ··· 114
　　5.6.3　多对多映射 ··· 116
5.7　动态查询 ··· 119
5.8　简单配置 ··· 123
本章小结 ··· 125

第6章　数据操作——使用 MyBatis ··························· 126

6.1　MyBatis 介绍 ··· 126
6.2　快速入门 ··· 127
6.3　MyBatis 基本元素 ·· 130
6.4　♯{}与${} ··· 131
6.5　结果映射 ··· 132
6.6　注解方式 ··· 134
6.7　动态 SQL ·· 136
6.8　几个重要配置 ·· 138
本章小结 ··· 140

第7章　连接关系数据库 ··· 142

7.1　简单介绍 ··· 142
7.2　连接 MySQL 数据库 ·· 142
7.3　连接 MariaDB 数据库 ·· 144
7.4　连接 SQL Server 数据库 ···································· 145
7.5　连接 Oracle 数据库 ·· 146
7.6　连接多数据库 ·· 147
本章小结 ··· 152

第8章　操作 MongoDB ·· 153

8.1　MongoDB 介绍及安装 ······································· 153
8.2　MongoDB 基本操作 ·· 155
8.3　Spring Boot 2 连接 MongoDB ····························· 158
8.4　使用 MongoTemplate 操作 ································· 159
8.5　使用 MongoRepository 接口操作 ························ 164
本章小结 ··· 168

第 9 章　Spring Boot 2 MVC … 169

9.1　MVC 介绍 … 169
9.2　配置 Maven … 170
9.3　建立模型 … 171
9.4　建立资源及服务 … 173
9.5　建立控制层 … 175
9.6　建立模板 … 176
9.7　系统配置 … 180
本章小结 … 181

第 10 章　Spring Boot 2 RESTful … 183

10.1　RESTful 介绍 … 183
10.2　Maven 相关配置 … 184
10.3　RESTful API 设计 … 185
10.4　Swagger 应用 … 189
10.5　RESTful API 测试工具 … 195
10.6　整合前端 … 196
本章小结 … 199

第 11 章　Spring Boot 2 安全 … 200

11.1　安全介绍 … 200
11.2　Spring Boot 2 中快速整合 Spring Security … 201
11.3　更改自动配置方式 … 202
11.4　自定义加密配置方式 … 203
11.5　使用 UserDetailsService … 205
11.6　使用 JDBC 认证方式 … 212
11.7　带前端认证 … 214
本章小结 … 217

第 12 章　Spring Boot 2 测试 … 218

12.1　JUnit 5 框架介绍 … 218
12.2　Spring Boot 2 集成 JUnit 5 … 219
12.3　JUnit 5 使用介绍 … 223
12.4　JUnit 5 完整示例 … 229
12.5　Maven 配置测试环境 … 236
本章小结 … 239

第 13 章 其他相关技术 ········· 240

13.1 上传文件 ········· 240
13.2 Lombok 应用 ········· 245
13.3 热部署 Devtools 应用 ········· 248
13.4 更改应用启动 Logo ········· 250
13.5 应用缓存 ········· 252
本章小结 ········· 256

附录 A　Maven 的使用 ········· 257

A.1 Maven 安装 ········· 257
A.2 Maven 配置 ········· 259
A.3 Maven 基本命令 ········· 260

附录 B　YAML 语法 ········· 262

B.1 转换工具命令 ········· 262
B.2 基本语法 ········· 263
B.2.1 对象表示法 ········· 263
B.2.2 数组表示法 ········· 264

附录 C　IDEA 工具介绍 ········· 265

附录 D　Tomcat 服务器 ········· 269

附录 E　本书源码的使用说明 ········· 272

第1章 初识Spring Boot 2

本章是全书的基础部分，将介绍微服务、Spring 和 Spring Boot 2 基本知识，通过 Spring 和 Spring Boot 2 的示例比较，一窥 Spring Boot 2 开发的敏捷性。同时，本章介绍 Spring Boot 2 中的 Maven POM 知识，以及开发 Spring Boot 2 的工具。

1.1 微服务介绍

当前，微服务比较热门，很多企业都开始从传统开发模式转向微服务模式开发。微服务可理解为将一些需要实现的较小功能集采用独立应用模式，以避免代码量过大和耦合度过高等问题，也即分而自治。传统开发模式，开发人员比较熟悉的是 MVC 开发模式，为了实现一个较小功能，需要整合一个很大的代码库，给后期维护同样带来很大不便。

在这样一种情况下，业界提出了微服务概念。微服务重在轻，能实现快速开发、测试和部署。这样，一个业务系统可以实现快速迭代，缩短周期。一般认为，微服务具有以下优点：

- 缩短生产时间；
- 较小配置；
- 易于部署；
- 简单的可扩展性；
- 与容器兼容。

下面是一个介绍微服务和传统开发架构的对比案例，一个打车软件的功能模块划分简单描述如图 1.1 所示。

这样的功能划分对于开发人员来说不难理解，但是不管怎样，比如采用 SOA 架构、MVC 模式等进行开发，最终该软件将打包成一个单体应用系统并被部署，那么，开发单体应用系统，这种模式比较适合于小项目，其优点是：

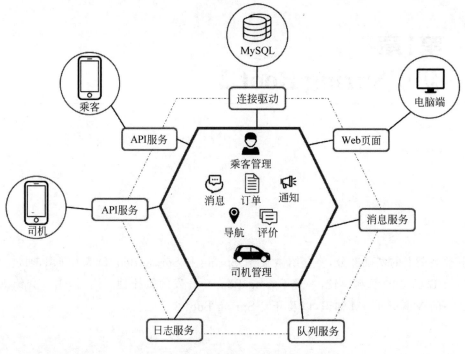

图 1.1 打车软件功能模块

- 集中式开发与管理；
- 代码复杂度不高；
- 模块间耦合度高；
- 部署简单。

这种开发模式比较适合于传统经典、小项目，业务模块能提前确定，开发周期较长，那么其缺点为：

- 开发维护难：由于代码耦合度较高，新入职员工学习成本高；
- 部署灵活差：由于是单体应用，每次更改哪怕一小段代码都需要重构，给部署带来一定复杂性，并且这个过程会比较长；
- 稳定性较差：单体应用系统可能会由于一个小的问题而崩溃；
- 开发协作差：如果由一个团队来进行开发，团队成员间改动代码，会造成沟通复杂，代码产生冲突可能性更高。

如果采用微服务架构，以上打车软件功能模块描述如图 1.2 所示。

可以将打车业务进一步细分为各个更小的微服务，服务间采用 REST 通信，其优点在此不再赘述。

本书介绍的 Spring Boot 就是微服务架构的一种实现，其可以大大简化开发模式，集成很多常用框架。为了对比 Spring Boot 开发快捷性，将和 Spring 开发进行简单对比，然后在后面各章节详细介绍 Spring Boot 实现。

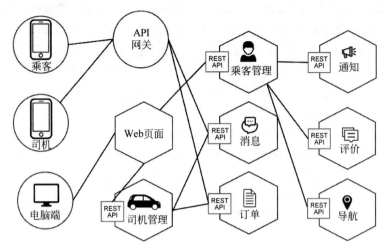

图 1.2 微服务架构示例

1.2 Spring 介绍

在开始 Spring Boot 之旅前，有必要先介绍 Spring 框架，毕竟先有 Spring，然后有 Spring Boot。随着互联网高度发展，应用系统越来越多，并且复杂性越来越大，开发周期逐步变长。为了解决这些问题，产生了各种能解决这些问题的框架，比如著名的 J2EE，但由于学习困难、开发效率低、代码重复高、配置复杂等原因，其在实践中并没有获得很好的成功。Spring 框架的出现是为了解决这些类似的问题，并且在多年的实践中，获得了业界认可，并广泛使用。

那么，Spring 是什么？Spring 是一个基于 Java 的开源框架，任何一个开发人员都可以愉快地使用该框架，正因如此，Spring 可以随时免费获得，并快速上手和使用，使其得到快速传播和壮大。

Spring 简化企业级应用系统开发，这是 Spring 诞生的初衷之一。在一个应用系统中，不同功能重复代码会很多，同时为了统一开发团队使用框架，降低学习复杂性，参考其他框架优缺点，Spring 具有简化企业级应用系统开发能力，并有效降低代码耦合度，极大地方便项目后期维护、升级和扩展等。而且，Spring 是一个 IOC(DI) 和 AOP 容器框架。

Spring 具有如下特点：
- 非侵入式。基于 Spring 开发的应用中的对象可以不依赖于 Spring 的 API；
- 控制反转。IOC(Inversion of Control) 指的是将对象的创建权交给 Spring 的 bean 工厂进行创建管理。这也是 Spring 的核心思想，通过面向接口编程的方式来实现对业务组件的动态依赖，IOC 是 Spring 主要解决程序的耦合方式。在开发中，Spring 通过配置文件说明要实例化的 java 类，然后通过加载方式读取配置文件，接着用 Spring 提供的方法获取根据指定配置进行初始化的实例对象；
- 依赖注入。DI(Dependency Injection) 是指将相互依赖的对象进行分离，在 Spring 的配置中描述它们之间的依赖关系，同时这些依赖关系只在使用时才会被建立；

- 面向切面编程。AOP(Aspect Oriented Programming)是一种编程思想,即OOP的延续。其思想是将系统中非核心的业务提取出来,单独进行处理;
- 容器。Spring是一个容器,管理对象的生命周期、对象与对象之间的依赖关系。也可以通过配置文件来定义对象,以及设置其他对象的依赖关系;
- 组件化。Spring通过使用简单的组件配置而组合成一个复杂的应用;
- 一站式。Spring提供开放、容器的方式,可以整合各种企业应用的开源框架和优秀的第三方类库。这也是Spring能广泛流行的原因之一。

Spring本身提供了多种模块供实际开发中选择,这样能够减轻开发中程序的复杂性,如图1.3所示。该图清楚描述了Spring内的多个模块,在系统开发中,这些模块提供了开发应用所需基本框架,但正如Spring解决紧耦合问题,不必将应用完全基于Spring框架,可以根据需要挑选适合的模块而忽略其余的模块。核心容器是最基本模块,定义了创建、配置和管理bean的方式,其余Spring模块都在其上进行构建。

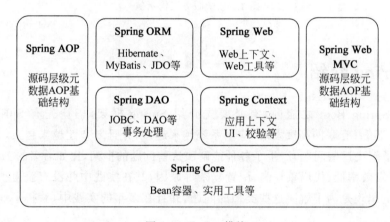

图1.3 Spring模块

组成Spring框架的每个模块都可以单独存在,或者与其他一个或多个模块共同实现系统。每个模块的功能描述如下:

核心容器。提供Spring框架的基本功能,主要组件是BeanFactory,其是工厂模式的实现。BeanFactory使用控制反转(IOC)模式将应用程序的配置和依赖性规范与实际的应用程序代码分开。

Spring上下文。是一个配置文件,向Spring框架提供上下文信息。其包括企业服务,例如JNDI、EJB、电子邮件、国际化、校验和调度功能。

Spring AOP。通过配置管理特性,该模块直接将面向切面的编程功能集成到Spring框架中。可以容易地使Spring框架管理的任何对象支持AOP。Spring AOP模块为基于Spring的应用程序中的对象提供了事务管理服务。通过使用Spring AOP,不用依赖EJB组件,就可以将声明性事务管理集成到应用程序中。

Spring DAO。JDBC DAO抽象层提供了有意义的异常层次结构,可用该结构来管理异常处理和不同数据库供应商抛出的错误消息。异常层次结构简化了错误处理,并且极大地降低了需要编写的异常代码数量(例如打开和关闭连接)。Spring DAO的面向JDBC的异常遵从通用的DAO异常层次结构。

Spring ORM。Spring 框架插入了若干个 ORM 框架，从而提供了 ORM 的对象关系工具，其中包括 JDO、Hibernate 和 iBatis SQL Map。所有这些都遵从 Spring 的通用事务和 DAO 异常层次结构。

Spring Web 模块。Web 上下文模块建立在应用程序上下文模块之上，为基于 Web 的应用程序提供了上下文。其还简化了处理多部分请求以及将请求参数绑定到域对象的工作，并提供了一些面向服务支持。

Spring MVC 框架。MVC 框架是一个全功能的构建 Web 应用程序的 MVC 实现。该框架可灵活配置，同时 MVC 容纳了大量视图技术，其中包括 JSP、Velocity、Tiles、iText 和 POI。

截至目前，Spring 版本已到 5.x，可以说，Spring 依然很受业界追捧，毕竟经过近 20 年的发展，Spring 逐渐壮大，内容更加丰富。

1.3 Spring 简单示例

鉴于目前使用 Spring 开发 Web 系统居多，本节将介绍如何通过 Spring 搭建一个简单的 Web 工程。通过本示例，可以一窥使用 Spring 开发 Web 应用的快捷和方便。本节涉及内容可以不做深究，后面会有详细介绍。当前，互联网管理 Java Jar 包方式较之前已发生较大变化，那么本书将采用当前较流行方式进行，即采用 Maven 方式加载项目所需 Jar 包，开发工具采用开源 Eclipse 工具，该工具能从互联网上方便获取，利于整书的讲解。

首先，在 Eclipse 的菜单中，选择 File→New→Project，在弹出窗口中，选择 Maven Project 项，然后单击 Next 按钮，如图 1.4 所示。

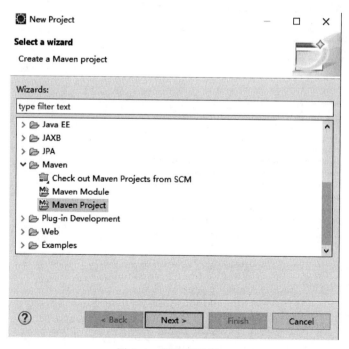

图 1.4 新建项目窗口

在新打开的窗口中，只勾选 Use default Workspace location，然后单击 Next 按钮，如图 1.5 所示。

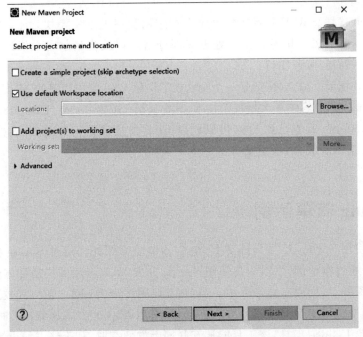

图 1.5　勾选 Use default Workspace location 窗口

在新打开窗口中，由于该示例创建的是 Web 项目，找到并选择 maven-archetype-webapp 项，然后单击 Next 按钮，如图 1.6 所示。

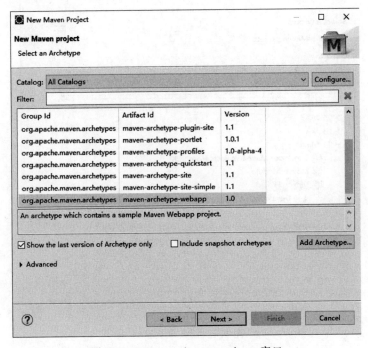

图 1.6　maven-archetype-webapp 窗口

在新打开窗口中，填写 Group Id 和 Artifact Id 项，填写内容示例如图 1.7 所示，然后单击 Finish 按钮，等待下载并自动配置完成。

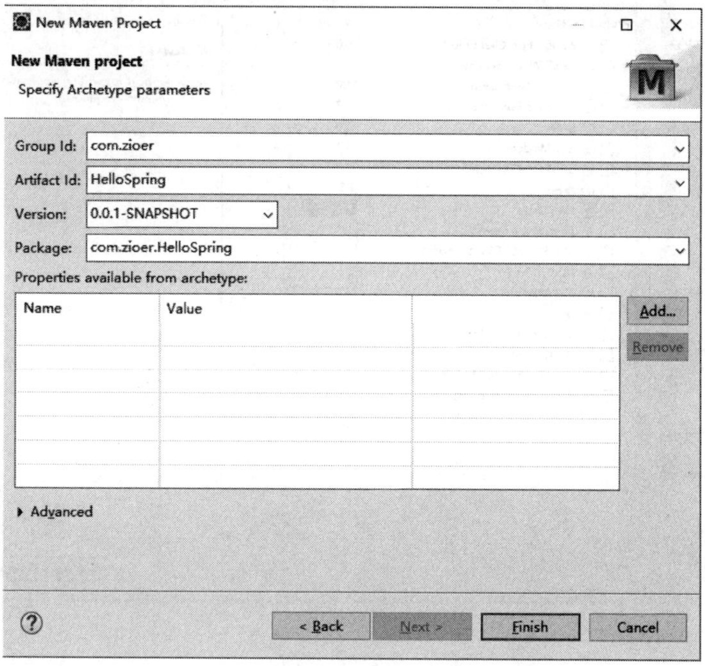

图 1.7　填写 Group Id 和 Artifact Id 窗口

等待完成，生成一个新的工程，结构如图 1.8 所示。

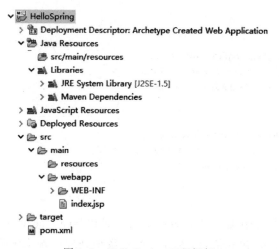

图 1.8　HelloSpring 工程框架

图 1.8 所示为一个空的框架工程，该工程还需要进行简单配置，才能完成基于 Spring Web 框架的搭建。由图 1.9 所示，生成的框架基于 Java 1.5，需要将其更改为 Java 1.8（本地安装 Java JDK 1.8），右键单击项目名称，在弹出菜单中选择 Properties，弹出修改属性窗口，如图 1.9 所示。

在图 1.9 中的左侧列表，选择 Project Facets，然后在右侧主窗口中修改 Java 1.5 版本

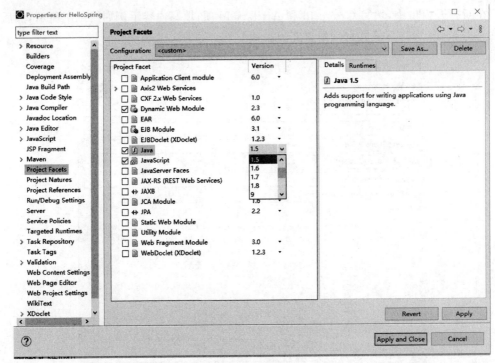

图 1.9 修改 Facets

为 1.8，然后单击 Apply and Close 按钮退出。

接着，打开项目中 pom.xml 文件，增加如下内容：

```xml
<dependencies>
    ...
    <dependency>
        <groupId>org.springframework</groupId>
        <artifactId>spring-webmvc</artifactId>
        <version>5.1.8.RELEASE</version>
    </dependency>
    <dependency>
        <groupId>javax.servlet</groupId>
        <artifactId>javax.servlet-api</artifactId>
        <version>3.1.0</version>
    </dependency>
</dependencies>
```

以上代码表示在项目中加入了 spring-webmvc 和 javax.servlet-api 依赖，然后保存该文件，项目将下载所需 Jar 包和依赖，等待下载和系统处理依赖完成后，再次打开项目依赖列表，如图 1.10 所示。

在图 1.10 中，项目所需 spring-webmvc 相关 Jar 包自动附加到工程中，省去开发人员手动加入各种 Jar 包的烦琐，项目所需 Jar 包加载完成。下面是工程配置。

在工程中，打开 web.xml 文件，加入下面内容：

```xml
<web-app>
    <servlet>
```

图1.10　Maven 依赖

```
    <servlet-name>dispatcher</servlet-name>
    <servlet-class>org.springframework.web.servlet.DispatcherServlet</servlet-class>
      <init-param>
        <!-- SpringMVC配置参数文件的位置 -->
        <param-name>contextConfigLocation</param-name>
        <!-- 默认名称为ServletName-servlet.xml -->
        <param-value>classpath:spring-*.xml</param-value>
      </init-param>
      <!-- 启动顺序,数字越小,启动越早 -->
      <load-on-startup>1</load-on-startup>
  </servlet>
  <!-- 所有请求都会被springmvc拦截 -->
  <servlet-mapping>
    <servlet-name>dispatcher</servlet-name>
    <url-pattern>/</url-pattern>
  </servlet-mapping>
  <context-param>
    <param-name>contextConfigLocation</param-name>
    <param-value>classpath:classpath:spring-*.xml</param-value>
  </context-param>
  <listener>
    <listener-class>org.springframework.web.context.ContextLoaderListener
</listener-class>
  </listener>
</web-app>
```

保存后,在路径 src\main\resources 下创建文件 spring-Context.xml,在该文件中增加如下内容：

```
<?xml version="1.0" encoding="UTF-8"?>
<beans xmlns="http://www.springframework.org/schema/beans"
    xmlns:context="http://www.springframework.org/schema/context"
    xmlns:xsi="http://www.w3.org/2001/XMLSchema-instance"
```

```xml
    xmlns:mvc = "http://www.springframework.org/schema/mvc"
    xsi:schemaLocation = "
        http://www.springframework.org/schema/beans
        http://www.springframework.org/schema/beans/spring-beans-4.3.xsd
        http://www.springframework.org/schema/context
        http://www.springframework.org/schema/context/spring-context-4.3.xsd
        http://www.springframework.org/schema/mvc
        http://www.springframework.org/schema/mvc/spring-mvc-4.3.xsd">

    <!-- 自动扫描包,实现支持注解的IOC -->
    <context:component-scan base-package = "controller"></context:component-scan>

    <!-- 支持mvc注解驱动 -->
    <mvc:annotation-driven />
    <!-- 视图页面配置 -->
    <bean class = "org.springframework.web.servlet.view.InternalResourceViewResolver">
        <property name = "prefix">
            <value>/WEB-INF/views/</value>
        </property>
        <property name = "suffix">
            <value>.jsp</value>
        </property>
    </bean>
</beans>
```

保存以上内容,完成 Spring 工程的配置。下面在路径 src\main\java\controller 中创建简单类文件 HelloController.java,如果该路径不存在,则需要手动创建,文件内容如下:

```java
package controller;

import org.springframework.ui.Model;
import org.springframework.web.bind.annotation.RequestMapping;

@Controller
@RequestMapping("/hello")
public class HelloController {

    @RequestMapping("")
    public String hello(Model model) {
        model.addAttribute("message", "Hello Spring MVC World!");
        return "hello";
    }
}
```

以上内容只有一个方法,保存该文件。下面创建视图文件,在路径 src\main\webapp\WEB-INF\views 中创建文件 hello.jsp,如果该路径不存在,则需要手动创建,内容如下所示:

```jsp
<%@ page isELIgnored = "false" language = "java" contentType = "text/html; charset = UTF-8"
    pageEncoding = "UTF-8" %>
<html>
```

```
< body >
< h2 align = "center"> $ {message}</h2 >
</body >
</html >
```

保存该文件,至此,简单基于 Spring MVC 工程便完成,完成后的工程目录如图 1.11 所示。

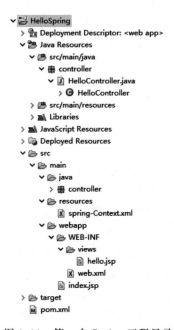

图 1.11 第一个 Spring 工程目录

将该工程部署到 tomcat 服务器,运行 tomcat 服务,浏览器打开下列网址:

http://127.0.0.1:8080/HelloSpring/hello/

访问得到图 1.12 所示内容,则表示成功。

图 1.12 访问网页

以上内容不多,但涉及很多的知识点,没有详细解释,在后面的章节中将会逐渐接触。通过该示例,可知 Spring 是很优秀的开源框架,并很容易搭建一个基于 MVC 框架的 Web 应用。

1.4　Spring Boot 介绍

既然 Spring 已经很优秀了,那为什么又产生了 Spring Boot 呢? Spring 经过多年的完善和发展,其已变得越来越重,而目前互联网应用逐渐向轻量化、快速化和敏捷化发展,显然在某些应用方面,Spring 显得不合时宜。此时,由 Spring 原创团队在 Spring 基础上开发了更加轻量级框架,即 Spring Boot。目前,最新版本已到 2.2.x。故本书介绍 Spring Boot 2 即指 2.x 版本系列,Spring Boot 2 更新很快,基本 1～2 月便有一个小版本发布。

比如,Spring 中配置文件和 Java 代码开发是两种不同语言,配置和开发分开,目前更倾向于将其混合,而不是开发人员要同时掌握多种开发语言。Spring Boot 基于 Spring 4.0 以上版本设计,不仅继承了 Spring 框架原有的优秀特性,还通过简化配置来进一步简化了 Spring 应用的整个搭建和开发过程。同时,Spring Boot 通过集成大量的框架使得依赖包的版本冲突,以及引用的不稳定性等问题也得到了很好的解决。

以下是 Spring Boot 的一些特点:

- 快速创建独立的 Spring 应用;
- 可嵌入 Tomcat、Jetty、Undertow 等,而且不需要单独进行部署;
- 提供的 starters poms 简化 Maven 配置;
- 尽可能自动配置 Spring 应用;
- 提供生产指标、健壮检查和外部化配置;
- 没有代码生成和 XML 配置要求。

Spring Boot 框架中有两个非常重要的策略,即开箱即用和约定优于配置。

开箱即用:指在开发过程中,通过在 Maven 项目的 pom 文件中添加相关依赖包,然后使用对应注解来代替烦琐的 XML 配置文件以管理对象的生命周期。这个特点使得开发人员摆脱了复杂的配置工作以及依赖的管理工作,使其更加专注于业务逻辑。

约定优于配置:是一种由 Spring Boot 本身来配置目标结构,由开发人员在结构中添加信息的软件设计范式。这一特点虽增加了 BUG 定位的复杂性,但降低了部分灵活性,减少了开发人员需要做出决定的数量,减少了大量的 XML 配置,并且可以将代码编译、测试和打包等工作自动化。

Spring Boot 具有如下优点:

- 支持快速开发出 RESTful 风格的微服务架构;
- 自动化、适合做微服务,单一 Jar 包部署和管理非常方便。只要系统架构设计合理,大型项目也能用,加上用 Nginx 负载均衡,能轻松实现横向扩展。

由此,Spring Boot 具有 Spring 所具有的优点和基本特性,更增加了其自己独特的特性。其要解决的问题,一方面是精简配置,另一方面是如何更加方便地让 Spring 产生的生态圈和其他工具有效整合(比如 Redis、Email、ElasticSsearch)。

1.5　Spring Boot 2 示例

本节将介绍一个基于 Spring Boot 2 的简单 Web 项目。首先,在 Eclipse 的菜单中,选择 File→New→Project,在弹出窗口中,选择 Maven Project 项,然后单击 Next 按钮。

在新打开的窗口中勾选第一项,即创建一个简单项目,然后单击 Next 按钮,打开一个新窗口,如图 1.13 所示。填写 Group Id 和 Artifact Id 项,然后单击 Finish 按钮,完成新项目的创建。

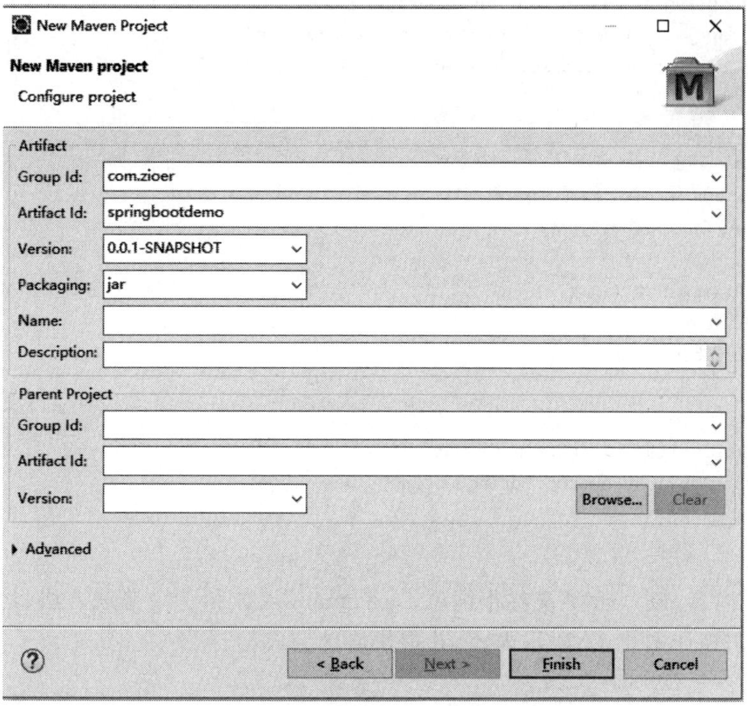

图 1.13　New Maven 项目窗口

在新建项目中打开 pom.xml 文件,增加如下内容:

```xml
<properties>
    <maven-jar-plugin.version>3.1.1</maven-jar-plugin.version>
</properties>
<!-- Spring Boot 父引用 -->
<parent>
    <groupId>org.springframework.boot</groupId>
    <artifactId>spring-boot-starter-parent</artifactId>
    <version>2.2.6.RELEASE</version>
</parent>
<dependencies>
    <dependency>
        <groupId>org.springframework.boot</groupId>
        <artifactId>spring-boot-starter-web</artifactId>
    </dependency>
</dependencies>
```

以上内容很简洁,元素 properties 中内容是为了解决 Eclipse 存在的一个 bug,但不影响项目运行。元素 parent 和 dependency 属于核心依赖,在这里使用了版本 2.2.6.RELEASE。保存该 pom 文件后,在路径 src\main\java\com\zioer\controller 创建文件 SpringBootApp.java,内容如下:

```java
package com.zioer.controller;

import org.springframework.boot.SpringApplication;
```

```java
import org.springframework.boot.autoconfigure.EnableAutoConfiguration;
import org.springframework.web.bind.annotation.RequestMapping;
import org.springframework.web.bind.annotation.RestController;

@RestController
@EnableAutoConfiguration
//该类装配Spring Boot内置Tomcat中
public class SpringBootApp {

    @RequestMapping("/index")
    public String test() {
        return "success";
    }

    public static void main(String[] args) {
        SpringApplication.run(SpringBootApp.class, args);
    }
}
```

保存以上内容，该java类文件很简单，一个main函数和一个普通方法，运行该文件，在Console窗口出现如图1.14所示内容，并很快启动完成。

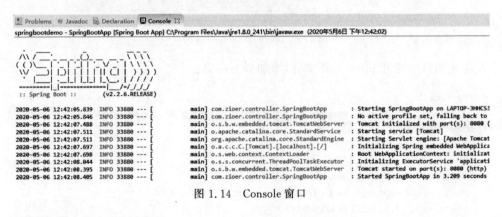

图1.14　Console窗口

在浏览器中输入下列网址：

$$http://127.0.0.1:8080/index$$

访问可得到如图1.15所示内容。

图1.15　访问Spring Boot简单项目

启动并运行成功。如图1.16所示为该项目工程目录。

该项目的Maven依赖如图1.17所示。

图1.17列出的只是其中一部分，实际上该项目依赖的Jar包很多。由以上示例可知，创建Spring Boot项目非常简单，很容易构建一个基本框架，并解决基于Spring Boot依赖问题，其重点在于（POM）文件中的starter依赖，给开发人员减负，开发人员无需关注太多东西，只需要关注其中一小部分内容即可。

图 1.16　springbootdemo 目录 　　　　图 1.17　项目依赖

实际上，Spring Boot 提供了很多这样类似的 starter 依赖，常见的有：

- spring-boot-starter-web；
- spring-boot-starter-data-jpa；
- spring-boot-starter-security；
- spring-boot-starter-test；
- spring-boot-starter-thymeleaf；
- spring-boot-starter-jdbc；
- spring-boot-starter-json。

本节介绍 Spring Boot 的一个简单示例，实现了一个简单功能，篇幅不多，由此可见，Spring Boot 实现一个开发应用系统基本框架，没有复杂的 XML 配置，需编写 pom 文件和一个 java 类，便能成功运行。

1.6　Maven POM 文件介绍

笔者之前写过一本书，由于 Jar 包管理工具还不成熟，需要手动加入 Jar 包，现在 Maven 已经很成熟，并被广泛应用，其作为管理 Jar 工具包很方便，当然还有其他的一些管理 Jar 工具，比如 Ant 和 Gradle。本书主要使用 Maven 构建工具，故在这里有必要用一节进行讲解。

Maven 是 Java 世界中构建工具之一，可以通过一小段描述信息来管理项目的构建。构建软件项目主要包括以下任务：下载依赖项，并在类路径上放置其他 Jar，将源代码编译为二进制代码，运行测试，然后将编译后的代码打包成可部署的工件（如 JAR、WAR 和 ZIP 文件），以及部署这些工件到应用程序服务器或存储库。Maven 能自动执行这些任务，最大限度地降低人工在手动构建软件时出错的可能，并将编译和打包代码的工作与代码构造分开。Maven 主要功能有依赖管理系统、多模块构建、一致的项目结构、一致的构建模型和插件机制。

Maven 项目的配置是通过项目对象模型（POM）完成的，即用 pom.xml 文件表示。POM 描述了项目、管理依赖项，并配置用于构建软件的插件。正如前面示例中 pom 文件展

示。下面是 Maven 基本结构:

```xml
<project xmlns="http://maven.apache.org/POM/4.0.0"
    xmlns:xsi="http://www.w3.org/2001/XMLSchema-instance"
    xsi:schemaLocation="http://maven.apache.org/POM/4.0.0
    http://maven.apache.org/xsd/maven-4.0.0.xsd">
    <modelVersion>4.0.0</modelVersion>
    <groupId>com.zioer</groupId>
    <artifactId>pomdeom</artifactId>
    <version>0.0.1-SNAPSHOT</version>
    <dependencies>
        <dependency>
            ...
        </dependency>
    </dependencies>
    <build>
        <plugins>
            <plugin>
                ...
            </plugin>
        </plugins>
    </build>
</project>
```

上面代码列出的是 POM 基本结构,其中元素 project 是所有 pom.xml 的根元素,它还声明了一些 POM 相关的命名空间及 xsd 元素等。

- modelVersion:用于描述 POM 模型的版本;
- groupId:用于描述创建项目所属公司或组唯一基本名称。这个名称应该和项目所在的组织或公司存在关联,定义方法一般都是网址倒序,例如 com.zioer;
- artifactId:用于描述当前项目的唯一名称;
- version:描述当前项目的版本,SNAPSHOT 意为快照,说明该项目还处在开发中,还有其他一些元素没有自动生成,比如下面元素;
- Packaging:用于描述打包方式,比如 war、jar 或 zip 等;
- Name:用于描述一个对于用户更为友好的项目名称;
- URL:用于描述项目的主页。

其中,groupId、artifactId 和 version 组合在一起便形成唯一标识符,其是指定项目将使用哪些外部库版本的原理。

Maven 的依赖管理是重点,项目无需在本地管理这些依赖,而是需要时从中央存储仓库中自动下载,那么在实际中,这就是依赖,需要提供 groupId、artifactId 和库的版本。下面是示例:

```xml
<dependency>
    <groupId>org.springframework</groupId>
    <artifactId>spring-core</artifactId>
    <version>5.1.8.RELEASE</version>
</dependency>
```

Maven 在处理依赖项时,它会将 Spring Core 库中指定版本下载到本地 Maven 存储库中,并在项目中引用。如果在依赖中没有指定版本,则会下载最新版本。

在 pom.xml 中,可以使用元素 properties 来管理各依赖的版本,有助于文档的阅读,比如下面示例:

```
<properties>
    <spring.version>5.1.8.RELEASE</spring.version>
</properties>
<dependencies>
    <dependency>
        <groupId>org.springframework</groupId>
        <artifactId>spring-core</artifactId>
        <version>${spring.version}</version>
    </dependency>
</dependencies>
```

这样,当有多个地方用到同一个元素 spring.version 时,只需要引用该元素即可,便于修改一处,达到统一修改的目的。

在 POM 文件中,一个重要的元素是 build,其提供了 Maven 目标、已编译项目的目录以及应用程序的最终名称的描述。基本形式如以下代码所示:

```
<build>
    <defaultGoal>install</defaultGoal>
    <directory>${basedir}/target</directory>
    <finalName>${artifactId}-${version}</finalName>
    <filters>
        <filter>filters/filter1.properties</filter>
    </filters>
    //...
</build>
```

在上面代码中,元素介绍如下:
- defaultGoal:执行 build 任务时,如果没有指定目标,将使用的默认值;
- directory:用于描述 build 目标文件的存放目录,默认在 ${basedir}/target 目录;
- finalName:build 目标文件的名称,默认为 ${artifactId}-${version};
- filters:定义 *.properties 文件,包含一个 properties 列表,该列表会应用到支持 filter 的 resources 中。

在 POM 文件中,另一个重要特性是它对配置文件的支持,即元素 profiles,包含了一组配置值。通过使用配置文件,可以为不同的环境(如生产/测试/开发)自定义构建,基本形式如以下代码所示:

```
<profiles>
    <profile>
        <id>production</id>
        <build>...</build>
    </profile>
</profiles>
```

以上介绍 POM 文件的基本构成,并能根据项目实际构建需要的 POM 文件,Maven 重在解决 Jar 包依赖,并为项目中 Jar 包版本升级、测试、打包等提供方便,更重要的是,开发人员只需要关注项目具体本身,不再需要为其各种依赖所干扰。

1.7 使用 Eclipse

工欲善其事,必先利其器。为了能高效、快速开发应用系统,必须有一款得心应手工具。Eclipse 是一个开源的集成开发工具,作为一款可高效开发 Java 的工具,重点是可以免费获得,值得推荐,本书将基于 Eclipse 进行讲解,本节重点介绍 Eclipse 的一些基本用法。

在 Eclipse 的官网下载最新版本,下载网址:

https://www.eclipse.org/downloads/

Eclipse 提供多平台版本下载,下载带压缩包形式,下载完成后,解压缩便可运行可执行,无须安装,如图 1.18 所示,最好将解压缩后的目录放到一个固定的目录下,在桌面建立快捷方式,便于快速启动。注意,运行之前,需要本机安装有 Java JDK 1.8 以上版本,本书所有示例基于 Java JDK 1.8。

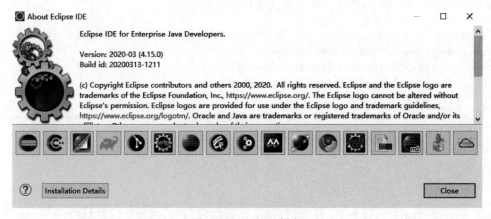

图 1.18 Eclipse 版本

初次进入后的主操作界面如图 1.19 所示。

Eclipse 操作界面比较简洁,最上面是常用菜单和快捷图标,左侧在首次运行显示快捷操作链接,一旦创建工程后,将显示工程结构,中间工作区域用于创建工程、编写代码等,工作区下方显示常用操作标签,或根据需要增加标签,比如 Servers 和 Markers 标签。在首次启动后,先别着急工作,重要的是进行全局设置,使之符合开发人员习惯。

单击 Window→Preferences,弹出属性窗口,在该窗口下可以进行一些常规设置,如图 1.20 所示。

在图 1.20 所示属性窗口中,左侧列表是所有能设置的项,有两种方式进行选择,一是在左侧列表一层层选择进行查找,然后单击所需项,另一种方式是在左侧上部输入框中输入属性名称,左边列表自动显示相关项,然后选择。当选择某一项后,右侧工作区将显示相关内容供编辑。

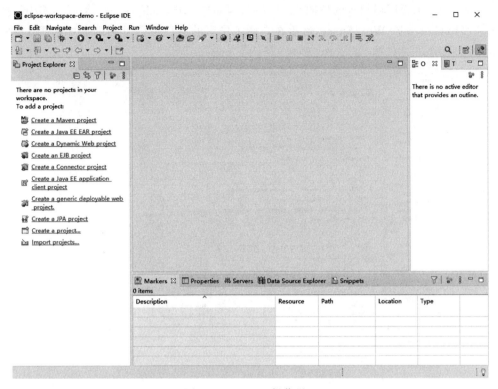

图 1.19　Eclipse 操作界面

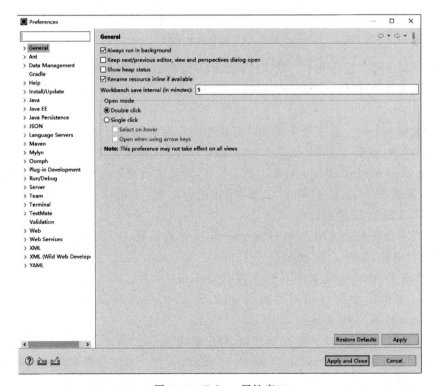

图 1.20　Eclipse 属性窗口

目前，Java 工程用得较多的编码是 UTF-8，但多数时候，Eclipse 默认编码是 GBK，修改方式是选择 General→WorkSpace，然后在右侧工作区的文件编码下列列表中选择 UTF-8，如图 1-21 所示。

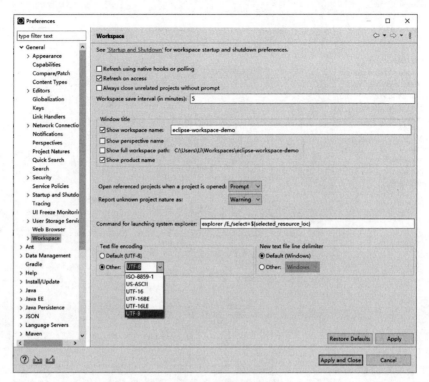

图 1.21　编辑文件编码

其次，编辑工作区字体大小。默认代码显示字体较小，根据个人习惯可修改主工作区字体大小，修改方式是选择 General→Appearance→Colors and Fonts，然后在右侧进行相关项操作。

第三，编辑 Maven。这主要考虑在计算机中可能存在多种开发工具，可能会共用同一个本地 Maven 库，或者 Eclipse 自带 Maven 比较低，需要更换更高版本的 Maven。在左侧列表中选择 Maven→Installations，在右侧工作区，默认显示的是内置的 Maven，如果需要使用本机自行安装的 Maven，则单击"Add"按钮进行添加，然后在列表中选择本地添加的 Maven 即可，如图 1.22 所示。

接着，在左侧列表选择 Maven→User Settings 进行设置，主要设置自定义 Maven 的配置文件及本地仓库，如图 1.23 所示。

以上完成了 Maven 设置。下面是进行 Servers 设置。Eclipse 默认没有内置安装 Server 服务，如果需要调试则需要进行相应设置，在左侧列表选择 Server→Runtime Environments，如图 1.24 所示。

图 1.24 默认为空，此时单击 Add 按钮，弹出新增 Server 窗口，如图 1.25 所示。Eclipse 支持多种应用服务器，包括 Tomcat、WebSphere 和 JBoss 等。目前业界常用的服务器是开源 Tomcat，下载最新版 Eclipse 会支持版本较高的 Tomcat，在这里选择 Tomcat v9.0，然后单

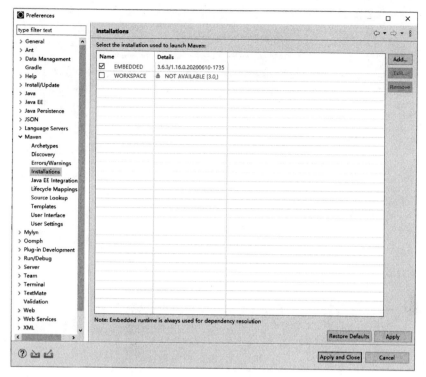

图1.22　Maven安装

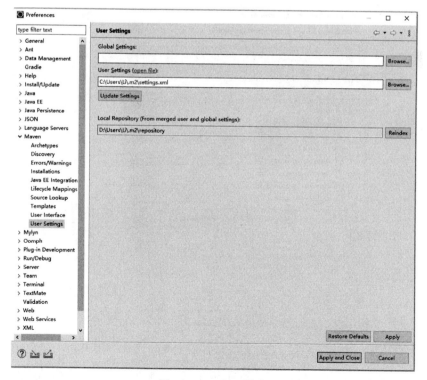

图1.23　Maven设置

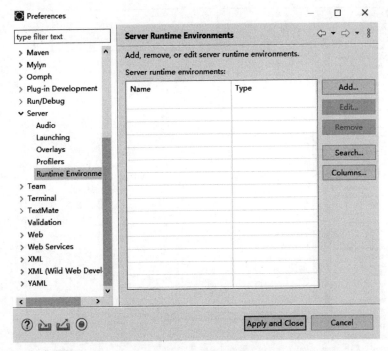

图 1.24　Server 安装

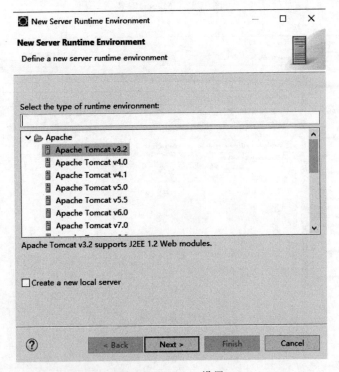

图 1.25　Server 设置

击 Next 按钮,在弹出窗口中选择本地安装的 Tomcat 服务器位置以及相应的 JDK,单击 Finish 按钮,完成 Server 配置。

以上设置完成后,在图 1.24 所示界面中单击 Apply and Close 完成所有设置,也可根据自己开发需要进行其他设置,或在以后开发中进行设置。回到 Eclipse 主界面后,在 Servers 选项卡中还是空的,单击快捷链接,如图 1.26 所示。

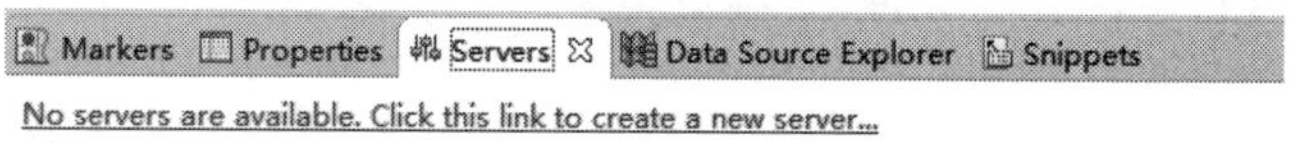

图 1.26　Servers 选项卡

在打开的建立新 Server 窗口中,选择刚新建的 Tomcat 服务器,并输入相应名称,单击 Finish 按钮,完成 Server 配置,此时 Servers 选项卡便出现 Tomcat 服务器,如图 1.27 所示。同时,在主界面左侧列表出现 Servers 项,在该项中,可以对 Tomcat 服务进行一些常用设置。

图 1.27　Tomcat 服务器

以上是对 Eclipse 中重点设置进行介绍,如何建立 Maven 应用,在前面已有相关介绍。还有一个常用操作是导入操作,位于菜单 File→Import,通过该菜单,可以导入已有工程,包括本书中示例工程。本节中涉及的 Maven 和 Tomcat 服务器的安装请详见附录。

1.8　使用 Spring Tool Suite 4

Spring Tool Suite 可理解为基于 Eclipse 等工具的一个套件,提供了一个已封装、可使用环境来快速生成、实现、调试、运行和部署 Spring Boot 应用,特别适合于想快速理解和使用 Spring Boot 2 的开发人员。目前最新版本是 4.x,支持生成基本 Spring Boot 2 项目框架。

有多种获得和安装 Spring Tool Suite 4 的方法,一种是直接从官网下载已经集成了 Spring Tool Suite 4 的 Eclipse,地址:

https://spring.io/tools

下载对应操作系统的版本,解压缩后即可使用,需要进行简单设置,方法同 1.7 节介绍。

如果开发人员本地已有了 Eclipse,不想重新下载,则在 Eclipse 中,选择菜单 Help→Install New Software,弹出安装新插件界面,根据 Eclipse 版本,输入下载地址,Eclipse 和 Spring Tool Suite 4 版本对应关系示例如表 1.1 所示。建议下载 Eclipse 的最新版本,体验性更好。

表 1.1　Eclipse 和 Spring Tool Suite 4 版本对应关系示例

Eclipse 版本	Spring Tool Suite4 版本
4.14	https://download.springsource.com/release/TOOLS/sts4/update/e4.14/
4.13	https://download.springsource.com/release/TOOLS/sts4/update/e4.13/
4.12	https://download.springsource.com/release/TOOLS/sts4/update/e4.12/

比如，本地 Eclipse 是 4.12，则在地址栏中输入：

https：//download.springsource.com/release/TOOLS/sts4/update/e4.12/

如图 1.28 所示，单击 Enter 键，等待下载列表，直到列表框出现能下载内容列表，选择全部，单击 Next 按钮，等待系统处理，直到出现下一个页面。

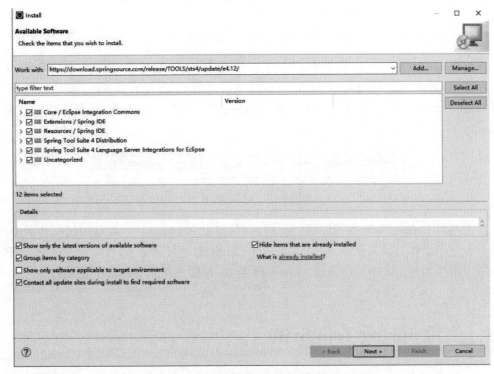

图 1.28　安装插件页面

然后，跟着页面一步步操作，直到结束。整个过程视网速情况而有所不同，断网情况不可操作。安装完成后，系统会提示重启软件。

以上两种方法都可获得 Spring Tool Suite 4 体验。当使用的是集成版时，启动界面如图 1.29 所示。

图 1.29　集成 Spring Tool Suite 4 启动界面

下面介绍在 Eclipse 中的简单操作方法。

选择菜单 File→New→Other，在新建窗口中选择 Spring Starter Project 项，单击 Next 按钮，如图 1.30 所示。

弹出详细设置页。在该页面填写内容较多，有很多项在前面介绍过，在这里还可选择

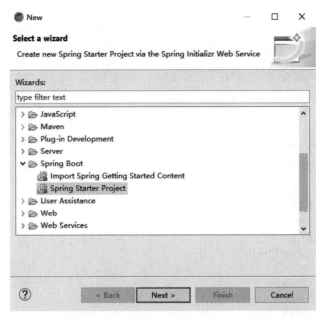

图 1.30　选择 Spring starter Project

Jar 包管理 Type 类型,默认是 Maven;包类型 Packaging 列表默认是 Jar,以及填写 Package 等,如图 1.31 所示。填写完成后,单击 Next 按钮。

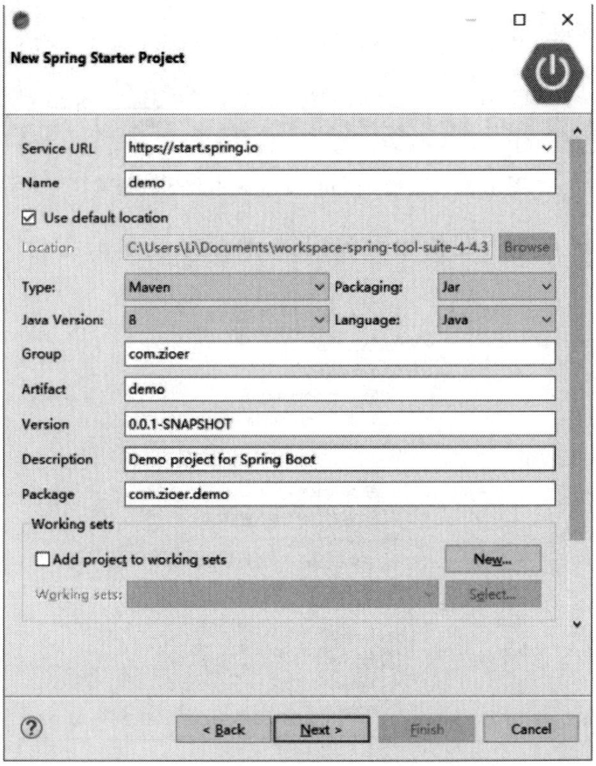

图 1.31　Spring Starter Project 详细页

接着，选择 Spring Boot 版本和加载依赖项，在这里默认 Spring Boot 版本是 2.2.6，选择 Spring Web Starter 后，单击 Finish 按钮，如图 1.32 所示。在该页面有一个比较人性化的小细节，即能自动记录经常使用的依赖，方便下次新建时快速选择。

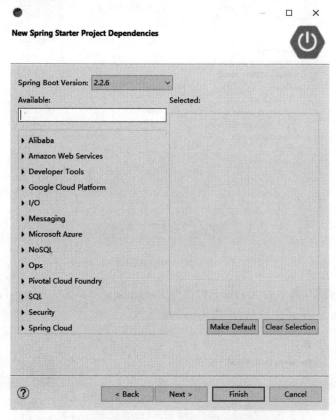

图 1.32　选择 Spring Boot 版本和加载依赖项

操作完成后，系统自动下载依赖项和配置，直到系统显示完成。生成的工程结构如图 1.33 所示。

图 1.33　生成的工程目录

这是一个标准的 Java Maven 工程,这个工程已能直接运行,右击该工程,在弹出的菜单中选择 Run As→Maven builder,弹出运行配置窗口,在 Goals 输入框中输入 spring-boot:run,单击 Run 按钮,运行结束后,Console 窗口中出现如图 1.34 所示内容,运行成功。

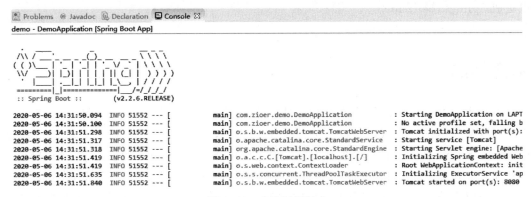

图 1.34　成功运行标志

打开 DemoApplication.java 文件,查看自动生成内容,如图 1.35 所示。

```
1  package com.zioer.demo;
2
3  import org.springframework.boot.SpringApplication;
4
5
6  @SpringBootApplication
7  public class DemoApplication {
8
9      public static void main(String[] args) {
10         SpringApplication.run(DemoApplication.class, args);
11     }
12
13 }
```

图 1.35　DemoApplication.java 文件

这样一个快速生成简单 Spring Boot 的工程模板,可让开发人员快速上手,尤其是前面的新建向导非常有用。Spring Tool Suite 4 还提供了一个非常有用的 Boot 仪表盘（Boot DashBoard）,如图 1.36 所示。

通过该仪表盘,可以快速启动、重启、标记和检查 Spring Boot 项目。如果该仪表盘没有在工作区,则选择菜单 Window→Show view→Other,弹出显示视图窗口,在搜索输入框中输入 boot,快速查找 Boot Dashboard,如图 1.37 所示。

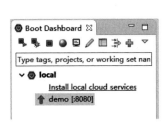

图 1.36　Boot 仪表盘

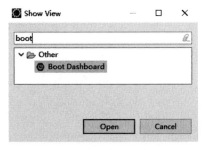

图 1.37　选择 Boot 仪表盘

选择 Boot Dashboard 项后,单击 Open 按钮,则工作区中将显示 Boot 仪表盘,便于操作。

如果开发人员没有使用 Eclipse,还可以选择 Visual Studio Code、Atom 等工具,同样可以集成 Spring Tool Suite 4。

本章小结

本章作为本书启动章节,重在介绍一些基本知识。先介绍了 Spring 项目,作者本人一直在开发基于 Spring 的项目,接触 Spring Boot 之后,为它的简洁所迷,所以,介绍了基于 Spring 和 Spring Boot 2 的两个实际案例。通过对比,Spring Boot 2 的创建、生成和启动变得更加简洁和方便;介绍开发 Spring Boot 2 的基本工具,包括 Maven、IDE 等,作为一个好的和得心应手的工具,会起到事半功倍的效果。为了更好理解 Maven,附录 A 中简单介绍了 Maven 相关内容。

第2章 启动Spring Boot 2

本章详细介绍开发 Spring Boot 2 的启动、依赖和配置，以及如何运行工程和打包工程。通过本章的学习，能掌握 Spring Boot 2 运行的基本方法，包含的重要类，以及如何运行。

2.1 启动类 MainApplication

在前面章节中，使用多种方法快速创建 Spring Boot 示例，无论从代码量，还是配置等方面，其已经简化了不少工作量。那么，Spring Boot 是如何启动的，从何处入口开启系统运行？

实际上，这得益于 Spring Boot 扫描机制，其在启动时，会扫描并找到注释@SpringBootApplication 和主要方法，然后开启运行，示例代码如下所示。

```
@SpringBootApplication
public class DemoApplication {
    public static void main(String[] args) {
        SpringApplication.run(DemoApplication.class, args);
    }
}
```

同时，该类需要包含静态方法 main()。在 Spring Boot 2 中，简化了注解，将注解@EnableAutoConfiguration、@ComponentScan 和 @SpringBootConfiguration 统一为@SpringBootApplication，这样代码更加简洁。查看@SpringBootApplication 源代码，如图 2.1 所示。

图 2.1 注解@SpringBootApplication

以上三个注解@EnableAutoConfiguration、@ComponentScan 和@SpringBootConfiguration 在实际开发中使用非常频繁，以至于源码作者将这三个注解进行简化操作，更便于开发者理解和使用。

注解在 Spring 体系中具有很重要的作用，同理，在 Spring Boot 体系中，注解也很重要，从启动开始便需要使用注解。在这里，通过注解@SpringBootApplication 达到多个目的，一是引导启动系统；其次是自动配置项目中添加的 Jar 依赖项，自动扫描所有 bean 和包声明。

最后，为了启动 Spring Boot 项目，需要在 main 方法中调用下面方法：

```
SpringApplication.run()
```

2.2 Spring Boot Starters

在 Java 开发程序中，经常比较头疼的是添加各种依赖，特别是依赖传递、版本号等问题。这些问题时刻困扰开发人员，在 Spring Boot 2 中，内置多种 Starter 依赖。在这里，Starter 可理解为启动器，其包含了一系列集成到应用里面的依赖包，可以快速集成 Spring 及其他技术，而不需要重复加入依赖包。

比如，spring-boot-starter-web 是用于编写 REST API 相关的依赖，比较常见，其 Maven 依赖写法如下：

```xml
<dependency>
    <groupId>org.springframework.boot</groupId>
    <artifactId>spring-boot-starter-web</artifactId>
</dependency>
```

当只包含以上依赖时，工程中加入的 Jar 包如图 2.2 所示。

图 2.2 spring-boot-starter-web 依赖

图 2.2 没有展示完所有 Jar 包依赖，但由此可见，其已自动加入项目需要的 Jar 包，不需要开发人员另外加入相同依赖 Jar 包。

Starter 的命名，如果是官方启动器，则是以 spring-boot-starter-开头命名，并被官方保留。第三方的启动器不能以 spring-boot 开头命名，建议名称：

{name}-spring-boot-starter

比如，常用的 mybatis 启动器命名为：

mybatis-spring-boot-starter

其 Maven 依赖书写方法：

```xml
<dependency>
    <groupId>org.mybatis.spring.boot</groupId>
    <artifactId>mybatis-spring-boot-starter</artifactId>
    <version>2.0.1</version>
</dependency>
```

在前面章节介绍过，Spring Boot 官方提供了一些开箱即用的启动类，开发人员可以根据需要增加其中一个或多个启动器，比如下面代码：

```xml
<dependencies>
    <dependency>
        <groupId>org.springframework.boot</groupId>
        <artifactId>spring-boot-starter</artifactId>
    </dependency>
    <dependency>
        <groupId>org.springframework</groupId>
        <artifactId>spring-aop</artifactId>
    </dependency>
    <dependency>
        <groupId>org.aspectj</groupId>
        <artifactId>aspectjweaver</artifactId>
    </dependency>
</dependencies>
```

以上依赖包含核心启动器、Spring AOP 和 AspectJ 启动器等相关依赖。

2.3 相关依赖

在上节介绍 Starter 启动依赖时，没有在依赖中书写出版本号，即如果在一个项目中，Starter 依赖较多时，每一个都写相同的版本号，不利于管理。而是将版本号书写在元素 parent 中。如下代码所示：

```xml
<parent>
    <groupId>org.springframework.boot</groupId>
    <artifactId>spring-boot-starter-parent</artifactId>
    <version>2.2.6.RELEASE</version>
    <relativePath/> <!-- lookup parent from repository -->
</parent>
```

所以，只需在元素 parent 中指定 Spring Boot 启动依赖项版本号，则在添加其他启动依赖项时，就无需指定 Spring Boot 版本号了。

一个完整的 pom.xml 文件示例如下代码：

```xml
<project xmlns="http://maven.apache.org/POM/4.0.0"
    xmlns:xsi="http://www.w3.org/2001/XMLSchema-instance"
    xsi:schemaLocation="http://maven.apache.org/POM/4.0.0
    http://maven.apache.org/xsd/maven-4.0.0.xsd">
    <modelVersion>4.0.0</modelVersion>
    <parent>
        <groupId>org.springframework.boot</groupId>
        <artifactId>spring-boot-starter-parent</artifactId>
        <version>2.2.6.RELEASE</version>
        <relativePath/> <!-- lookup parent from repository -->
    </parent>
    <groupId>com.zioer</groupId>
    <artifactId>c2-1</artifactId>
    <version>0.0.1-SNAPSHOT</version>
    <name>c2-1</name>

    <properties>
        <java.version>1.8</java.version>
    </properties>

    <dependencies>
        <dependency>
            <groupId>org.springframework.boot</groupId>
            <artifactId>spring-boot-starter-web</artifactId>
        </dependency>
        <dependency>
            <groupId>org.springframework.boot</groupId>
            <artifactId>spring-boot-starter-test</artifactId>
            <scope>test</scope>
        </dependency>
    </dependencies>
</project>
```

通过以上简单的 pom.xml 配置，完成在项目中加入所需的 Jar 依赖。

2.4 配置文件

创建 Spring Boot 项目后，在 resources 目录下生成一个空的 application.properties 配置文件，如果没有，可手动在该目录下创建该文件，Spring Boot 启动时会加载该配置文件。

该配置文件包含系统属性、环境变量、命令参数等信息。比如，内置 Tomcat，默认端口号是 8080，如果想更改启动端口号，则需要在该文件中配置不同端口号，如下代码所示：

```
server.port=8090
```

以上配置将项目启动时端口号更改为 8090。Properties 文件格式形如上面代码所示，以点号分隔各段，然后符号=后表示具体值。这种配置文件格式比较常见。同时，Spring Boot 支持另一个配置文件格式，即 YAML 语法。只需要将 application.properties 更改为 application.yml，语法格式发生改变，示例代码如下：

```
server:
  port: 8090
```

同样达到更改端口号目的。

YAML 语法基本格式是：

```
k: v
```

表示键值对，冒号后有空格。以空格的缩进来控制层级关系；只要是左对齐的一列数据，都是同一个层级的，属性和值是大小写敏感的。

比较以上两种配置的区别，YAML 较 properties 而言，其语法更加简洁，比如同一个层级名称只出现一次，而 properties 每次都将重复书写，故 YAML 文件将更加小巧，详细介绍见附录 B。从使用效果上比较，都能达到同样目的，即在使用上没有孰优孰劣，主要看开发人员使用习惯。

2.5　@RestController 和 @Value

注解 @RestController 在开发 Rest 服务时比较常用，可理解为直接返回内容。该注解实际是注解 @ResponseBody 和 @Controller 的合体，查看 @RestController 源码，如图 2.3 所示。

```
@Controller
@ResponseBody
public @interface RestController {
```

图 2.3　注解 @RestController

如果在一个类上有注解 @RestController，则该类中的方法无法返回具体页面，比如 jsp、html 或模板等，返回的将是 Return 里的内容。即如果希望返回一个页面，则不能采用注解 @RestController。

实际上，采用注解 @RestController，简化了代码工作量，直接明了地声明该类，很适合做 Rest 服务。

注解 @Value 用于在代码中读取配置文件中的属性值，书写方法：

@Value("${key}")

比如，在配置文件中有如下设置：

server.applicatonName = testApplication

在代码中，读取方法：

@Value("${server.applicatonName}")
String name;

采用以上方法可以方便读取出配置文件中的属性值，但是如果指定的属性不存在，则系

统在启动时会报下面错误:

Could not resolve placeholder 'server.applicatonName' in value "${server.applicatonName}"

为了解决以上属性可能不存在的问题,可以在注解@Value("${key}")中加入初始值,比如:

@Value("${key:defaultValue}")

为了更好理解以上注解,下面是详细示例代码:

```
@RestController
public class firstController {

    @Value("${server.applicatonName:systemName}")
    String name;

    @GetMapping("first")
    public List<Student> first() {
        List<Student> list = new ArrayList<Student>();
        Student stu = new Student();
        stu.setName(name);
        stu.setAge(12);
        stu.setGrade("first");

        list.add(stu);

        return list;
    }
}
```

在以上代码中,使用注解@Value读取配置文件中的一个属性值,并赋一个初始值,类Student是个自定义类,代码如下所示:

```
public class Student {
    private String name;
    private int age;
    private String grade;

    public String getName() {
        return name;
    }
    public void setName(String name) {
        this.name = name;
    }
    public int getAge() {
        return age;
    }
    public void setAge(int age) {
        this.age = age;
    }
    public String getGrade() {
```

```
        return grade;
    }
    public void setGrade(String grade) {
        this.grade = grade;
    }
}
```

2.6 logging

日志对于调试、错误记录具有很重要的作用。Spring Boot 默认使用 Logback 记录日志。如果使用了依赖 Spring Boot Starters，则无须再重复添加 Logback 依赖。此时，在没有额外的配置时，Logback 默认是将日志打印到控制台上。下面是如何使用日志的简单方法示例：

```
@Controller
public class DemoController {
    private static final Logger LOG = LoggerFactory.getLogger(DemoController.class);

    @GetMapping("demo")
    public String doThing() {
        LOG.info("log-info");
        LOG.warn("log-warn");
        LOG.debug("log-debug");
        LOG.trace("log-trace");
        LOG.error("log-error,{}","错误原因");

        return "OK";
    }
}
```

在上面的示例中，首先全局定义一个静态 Logger 变量，供该类中所有方法调用，在上面定义的方法 doThing 中，示例了 Logger 的多种级别的日志，包括 info、warn、debug、trace 和 error，当在前端页面中访问 demo 路径时，将在 Console 面板中打印出相关日志，如图 2.4 所示。

```
2019-11-11 19:03:59.501  INFO 34520 --- [nio-8091-exec-7] c.zioer.demo.controller.DemoController   : log-info
2019-11-11 19:03:59.501  WARN 34520 --- [nio-8091-exec-7] c.zioer.demo.controller.DemoController   : log-warn
2019-11-11 19:03:59.501 ERROR 34520 --- [nio-8091-exec-7] c.zioer.demo.controller.DemoController   : log-error,错误原因
```

图 2.4 日志输出

在图 2.4 中，日志显示格式从左至右显示内容如下所示：
- 日期和时间：运行时时间；
- 日志级别：包括 INFO、ERROR 和 WARN；
- 进程 ID 号；
- 线程名及内容等。

由此可知，在程序中集成日志管理是非常方便的，并且提供了多种日志级别供选择。但

默认是在 Console 面板或运行界面直接输出，显然不够实用。在 Spring Boot 2 中，提供了两种配置日志的方法：一是直接在配置文件 application.properties 中进行配置；另一种是采用单独的配置文件。

下面是在配置文件 application.properties 中配置的方法：

```
logging.pattern.console=%d{yyyy-MM-dd HH:mm:ss} %-5level %logger - %msg%n
```

以上配置用于更改 Console 显示输出格式，示例如图 2.5 所示。

```
2019-11-11 19:14:13  INFO   com.zioer.demo.controller.DemoController- log-info
2019-11-11 19:14:13  WARN   com.zioer.demo.controller.DemoController- log-warn
2019-11-11 19:14:13  ERROR  com.zioer.demo.controller.DemoController- log-error,错误原因
```

图 2.5　自定义 log 输出日志

同理，如果需要更改日志文件中的格式，配置示例代码如下：

```
logging.pattern.file=%d{yyyy-MM-dd HH:mm:ss} %-5level %logger - %msg%n
```

在上面代码中，重要的参数描述如下所示：

- %d{HH:mm:ss.SSS}：表示日志输出时间；
- %thread：表示输出日志的进程，在上面示例中省略；
- %-5level：日志级别，表示使用 5 个字符靠左对齐；
- %logger：日志输出项的名称；
- %msg：日志内容；
- %n：换行符。

如果希望日志记录到指定文件，有两种配置方法，一是直接指定日志输出位置和文件名，示例代码如下：

```
logging.file=d:/log/logging.log
```

另一种配置方法是只指定日志输出位置，示例代码如下所示：

```
logging.path=d:/log
```

以上配置只指定日志位置，在程序运行后将自动生成日志文件 spring.log。以上两个配置建议不要同时配置，否则，程序运行后将采用 logging.file 方式，而忽略另外一个。

上面介绍的是在配置文件 application.properties 中配置的简单方法，另一种推荐的配置方式是采用独立配置文件进行配置，该配置文件建议的文件名为 logback-spring.xml，可直接放置到 resources 目录下，程序启动后，将自动识别。下面是简单示例：

```
<?xml version="1.0" encoding="UTF-8"?>
<configuration>

    <property name="HOME_LOG" value="d:\\log\\logging.log"/>

    <appender name="FILE-ROLLING" class="ch.qos.logback.core.rolling.RollingFileAppender">
        <file>${HOME_LOG}</file>

        <rollingPolicy class="ch.qos.logback.core.rolling.SizeAndTimeBasedRollingPolicy">
```

```xml
<fileNamePattern>d:\\log\\app.%d{yyyy-MM-dd}.%i.log.gz</fileNamePattern>
            <!-- 设置每一个文件最大 10MB -->
            <maxFileSize>10MB</maxFileSize>
            <!-- 设置所有存档文件的总大小,如果总大小>10GB,它将删除旧的存档文件 -->
            <totalSizeCap>10GB</totalSizeCap>
            <!-- 保留最大天数 40 天 -->
            <maxHistory>40</maxHistory>
        </rollingPolicy>

        <encoder>
            <pattern>%d %p %c{1.} [%t] %m%n</pattern>
        </encoder>
    </appender>

    <logger name="com.zioer" level="WARN" additivity="false">
        <appender-ref ref="FILE-ROLLING"/>
    </logger>

    <root level="error">
        <appender-ref ref="FILE-ROLLING"/>
    </root>

</configuration>
```

以上自定义配置用于配置日志输出到文件,一个关键元素是 appender,该元素中的内容描述了日志文件增长和处理的方式,包括大小、归档和历史天数等内容。在日志配置文件中可以定义多个 appender 元素,可以对于不同环境进行描述,比如 Console 输出的描述,示例如下:

```xml
<appender name="CONSOLE" class="ch.qos.logback.core.ConsoleAppender">
    <layout class="ch.qos.logback.classic.PatternLayout">
        <Pattern>
            %d{HH:mm:ss.SSS} [%t] %-5level %logger{36} - %msg%n
        </Pattern>
    </layout>
</appender>
```

元素 logger 用于描述日志输出的层级,以及应用哪一个 appender,比如 Console 输出的指定,示例代码如下所示:

```xml
<logger name="com.zioer" level="debug" additivity="false">
    <appender-ref ref="CONSOLE"/>
</logger>
```

其中,属性 additivity 为 false,表示此 logger 的打印信息不向其上级传递。元素 root 的打印级别设置为 error,指定了名字为 FILE-ROLLING 的 appender。完整代码请查看源码。

以上介绍了两种常用的设置方法,并且简单可行。

2.7 运行工程

本节示例一个完整 Spring Boot 2 工程，项目结构如图 2.6 所示。

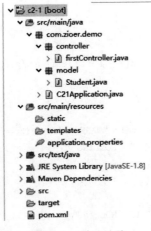

图 2.6　项目目录

如图 2.6 所示为一个比较典型的 Java 工程目录，包含了几个重要的目录：
- src/main/java：主要代码存放目录；
- src/main/java/controller：放置控制类文件；
- src/main/java/model：放置业务数据模型；
- src/main/resources：资源目录，存放配置文件、静态文件、模板等；
- src/test/java：存放测试文件。

以上工程建立完成后，下面是运行工程。运行该工程有多种方法。一是在 Boot 仪表盘选择该 Spring Boot 工程，单击运行按钮，或右键单击该工程，在弹出菜单中选择"(Re)start"命令，如图 2.7 所示。

启动完成后，该工程图标有点小变化，并标出其端口号，如图 2.8 所示。

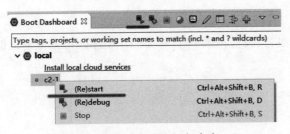

图 2.7　Boot 仪表盘运行命令

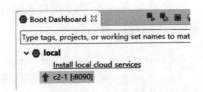

图 2.8　运行后的工程

其次，如果没有安装 Spring Tool Suite 4，则右击启动类 C21Application，在弹出菜单中选择 Run As→Java Application 命令，同样可以运行该工程。

第三，在 Package Explorer 面板中，右键单击该工程，在弹出菜单中选择 Run As→

Maven build 命令。首次运行时,将弹出配置窗口,在该窗口中的 Goals 输入框中输入:

spring-boot:run

如图 2.9 所示。然后单击 Run 按钮,运行该工程。

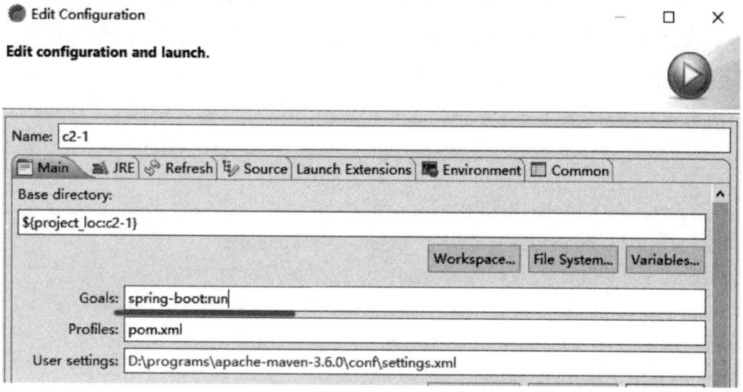

图 2.9 填写运行配置

通过以上多种方法,都可以正常启动该工程,然后在浏览器地址栏中输入下面地址访问:

http://127.0.0.1:8090/first

页面显示如图 2.10 所示。

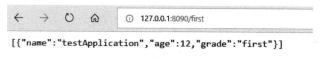

图 2.10 页面显示结果

出现如图 2.10 所示界面,表示整个工程正确无误。

2.8 打包工程

以上工程在 IDE 中可以正常运行,更重要的是需要将该工程打包,并能在实际环境中运行。一种简单可行的方法是,将其打包为一个可执行 Jar 包,这样,可以脱离类似于 Tomcat 运行环境运行。

在打包之前,确认 pom.xml 文件中有如下元素:

<packaging>jar</packaging>

以及

<build>
  <plugins>
    <plugin>
      <groupId>org.springframework.boot</groupId>

```xml
            <artifactId>spring-boot-maven-plugin</artifactId>
        </plugin>
    </plugins>
</build>
```

Spring Boot Maven plugin 能够将 Spring Boot 应用打包为可执行的 jar 或 war 文件，packaging 为 jar 表示将应用打包为 Jar 包，然后以通常的方式运行 Spring Boot 应用。以上 Maven 配置很重要，否则将打包不成功。

完整的 pom.xml 示例如下所示：

```xml
<project
    xmlns=http://maven.apache.org/POM/4.0.0
    xmlns:xsi="http://www.w3.org/2001/XMLSchema-instance"
    xsi:schemaLocation="http://maven.apache.org/POM/4.0.0
    http://maven.apache.org/xsd/maven-4.0.0.xsd">
    <modelVersion>4.0.0</modelVersion>
    <parent>
        <groupId>org.springframework.boot</groupId>
        <artifactId>spring-boot-starter-parent</artifactId>
        <version>2.2.6.RELEASE</version>
        <relativePath/> <!-- lookup parent from repository -->
    </parent>
    <groupId>com.zioer</groupId>
    <artifactId>c2-1</artifactId>
    <version>0.0.1-SNAPSHOT</version>
    <packaging>jar</packaging>
    <name>c2-1</name>

    <properties>
        <java.version>1.8</java.version>
    </properties>

    <dependencies>
        <dependency>
            <groupId>org.springframework.boot</groupId>
            <artifactId>spring-boot-starter-web</artifactId>
        </dependency>
        <dependency>
            <groupId>org.springframework.boot</groupId>
            <artifactId>spring-boot-starter-test</artifactId>
            <scope>test</scope>
        </dependency>
    </dependencies>

    <build>
        <plugins>
            <plugin>
                <groupId>org.springframework.boot</groupId>
                <artifactId>spring-boot-maven-plugin</artifactId>
            </plugin>
```

```
        </plugins>
    </build>
</project>
```

保存以上 pom.xml 文件，在 Package Explorer 面板中，右击该工程，在弹出的菜单中选择 Run As→Maven Clean 命令，等待命令结束。接着，右键单击该工程，在弹出的菜单中选择 Run As→Maven Install 命令，等待命令完成。打开工程目录，在 target 目录下生成该工程的 jar 文件，在命令提示符下查看，如图 2.11 所示。

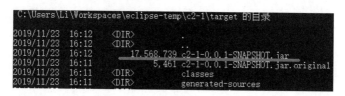

图 2.11　生成 jar 工程

运行该工程的方式是，在命令行输入下面命令：

java -jar c2-1-0.0.1-SNAPSHOT.jar

运行结果如图 2.12 所示。

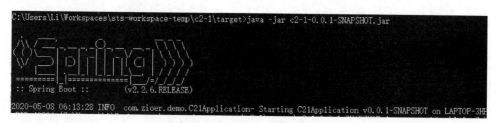

图 2.12　命令行运行 jar 包

本章小结

本章知识点较多，基本覆盖了从创建到最后生成可执行文件的全生命周期，内容包括启动类、Spring Boot 启动器、配置文件、示例注解、日志、运行工程和打包工程等，这个过程在系统开发中都会涉及。结合本章完整示例代码进行学习，效率会更高。

第3章 页面处理

本章介绍 Spring Boot 2 中页面处理支持的多种技术,包括 Thymeleaf、FreeMarker 和 JSP 等。

3.1 Thymeleaf 介绍

Thymeleaf 是用于 Web 和独立环境的现代服务器端 Java 模板引擎,为 Spring Boot 官方所推荐使用。其具有以下几个优点:
- Thymeleaf 可以静态地运行,支持 html 原型,然后在 html 标签里增加额外的属性来达到模板+数据的展示方式,既可以让美工在浏览器查看页面的静态效果,也可以让程序员在服务器查看带数据的动态页面效果;
- Thymeleaf 开箱即用。其提供标准和 Spring 标准两种方言,可以直接套用模板实现 JSTL、OGNL 表达式效果,避免每天套模板、修改 jstl 和标签的困扰。同时开发人员也可以扩展和创建自定义的方言;
- Thymeleaf 提供 Spring 标准方言和一个与 SpringMVC 完美集成的可选模块,可以快速地实现表单绑定、属性编辑器、国际化等功能。

Thymeleaf 支持处理 6 种模板,每种模板都称为模板模式:
- HTML。允许任何类型的 HTML 输入,包括 HTML5、HTML4 和 XHTML。将不会执行验证或格式检查,并且在输出中尽可能地遵守模板代码/结构;
- XML。允许 XML 输入。因此,代码应该是格式良好的——即没有未封闭的标签,没有未加引号的属性,等等,如果发现格式错误,解析器将会抛出异常;
- TEXT。允许对非标记性质的模板使用特殊语法。这种模板的例子可能是文本电子邮件或模板文档;
- JAVASCRIPT。允许处理 Thymeleaf 应用程序中的 JavaScript 文件。这意味着能够像在 HTML 文件中一样使用 JavaScript 文件中的模型数据。JAVASCRIPT 模板模式被认为是文本模式,因此使用与 TEXT 模板模式相同的特殊语法;

- CSS。允许处理 Thymeleaf 应用程序中涉及的 CSS 文件。类似于 JAVASCRIPT 模式，CSS 模板模式也是文本模式，并使用 TEXT 模板模式中的特殊处理语法；
- RAW。用于将未触及的资源（文件、URL 响应等）插入正在处理的模板中。例如，可以将 HTML 格式的外部非受控资源包含在应用程序模板中，从而安全地知道这些资源可能包含的任何 Thymeleaf 代码都不会被执行。

Thymeleaf 的目的是将优雅的自然模板集成到开发工作流程中，即 HTML 能够在浏览器中正确显示，并且可以作为静态原型。其次是提供一个优雅和可维护的创建模板方式，Thymeleaf 建立在自然模板的概念之上，以不影响模板作为设计原型的方式将其逻辑注入模板文件中，极大地改善了设计沟通，弥合了前端设计和开发人员之间的理解偏差。

3.2 集成 Thymeleaf

在创建的 Spring Boot 2 工程中要使用 Thymeleaf 模板，需要将 Thymeleaf 依赖加入工程的 pom.xml 文件中，如下代码所示：

```xml
<dependency>
    <groupId>org.springframework.boot</groupId>
    <artifactId>spring-boot-starter-thymeleaf</artifactId>
</dependency>
```

完整的 pom.xml 文件示例代码如下所示：

```xml
<project
xmlns="http://maven.apache.org/POM/4.0.0"
xmlns:xsi="http://www.w3.org/2001/XMLSchema-instance"
xsi:schemaLocation="http://maven.apache.org/POM/4.0.0
http://maven.apache.org/xsd/maven-4.0.0.xsd">
    <modelVersion>4.0.0</modelVersion>
    <parent>
        <groupId>org.springframework.boot</groupId>
        <artifactId>spring-boot-starter-parent</artifactId>
        <version>2.2.6.RELEASE</version>
        <relativePath/> <!-- lookup parent from repository -->
    </parent>
    <groupId>com.zioer</groupId>
    <artifactId>c3-1</artifactId>
    <version>0.0.1-SNAPSHOT</version>
    <packaging>jar</packaging>
    <name>c3-1</name>
    <description>demo</description>

    <properties>
        <java.version>1.8</java.version>
    </properties>

    <dependencies>
```

```xml
<dependency>
    <groupId>org.springframework.boot</groupId>
    <artifactId>spring-boot-starter-thymeleaf</artifactId>
</dependency>
<dependency>
    <groupId>org.springframework.boot</groupId>
    <artifactId>spring-boot-starter-web</artifactId>
</dependency>
<dependency>
    <groupId>org.springframework.boot</groupId>
    <artifactId>spring-boot-starter-test</artifactId>
    <scope>test</scope>
</dependency>
    </dependencies>

    <build>
        <plugins>
            <plugin>
                <groupId>org.springframework.boot</groupId>
                <artifactId>spring-boot-maven-plugin</artifactId>
            </plugin>
        </plugins>
    </build>
</project>
```

保存以上 pom.xml 文件,创建 Controller 类文件,示例代码如下:

```java
package com.zioer.demo.controller;

import org.springframework.stereotype.Controller;
import org.springframework.web.bind.annotation.GetMapping;

@Controller
public class mainController {
    @GetMapping("/main")
    public String main(Model model) {
        model.addAttribute("name","Zioer");
        return "main";
    }
}
```

以上代码中,使用了注解@Controller,故该类是一个控制类,方法 main()返回了一个字符串,该字符串表示返回视图名称。下面创建模板,该模板默认需放置到路径 src/main/resources/templates 下,文件名为 main.html,示例代码如下:

```html
<!DOCTYPE html>
<html>
<head>
    <meta charset="UTF-8" />
    <title>demo</title>
</head>
```

```
<body>
    <p th:text = "'Welcome : ' + ${name}"/>
</body>
</html>
```

以上模板文件较简单,通过 p 标签中的 th:text 属性,即 Thymeleaf 的语法,显示了后端传递值。保存以上模板文件,然后运行该工程,在浏览器中输入下列地址进行访问:

$$http://127.0.0.1:8080/main$$

访问页面结果如图 3.1 所示。

创建完成的工程目录结构如图 3.2 所示。

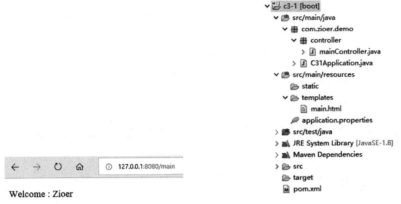

图 3.1 简单 Thymeleaf 工程　　图 3.2 简单 Thymeleaf 工程目录结构

提示:Thymeleaf 模板默认放置到 templates 目录下,其余静态文件,包括 css、js 等,需放置到 static 目录下。

3.3 Thymeleaf 语法

开箱即用的特点,是指 Thymeleaf 的核心库提供了一种"标准方言",对于快速开发非常有帮助。标准方言的多数处理器是属性处理器,允许浏览器甚至能在被处理之前正确显示 HTML 模板文件。例如下面代码所示:

```
<input type = "text" name = "userName" th:value = "${name}" />
```

以上示例可以被浏览器正确显示,而且还允许(可选地)在浏览器中静态打开原型时指定一个值属性,最终被解释时,在处理模板期间被 ${name} 传递值所取代。

这种方式将有助于设计人员和开发人员使用完全相同的模板文件,能有效减少将静态原型转换为工作模板文件所需的工作量。这种方式简单理解为自然模板。

本节将详细介绍 Thymeleaf 语法,以能更好理解、掌握和灵活运用该语法。

3.3.1 表达式语法

下面是 Thymeleaf 常用的表达式语法:

- 变量表达式：${...}

例如，${name}，${city.code}

- 选择变量表达式：*{...}

例如，通过 th：object 方式获取对象，然后使用 th：xx = "*{}"获取其中对象属性。

- 消息表达式：#{...}

国际化属性。

- 链接网址表达式：@{...}

实际用于拼接静态资源路径。

- 片段表达式：~{...}

~{viewName} 表示引入完整页面；~{viewName::selector} 表示在指定页面寻找片段，其中 selector 可为片段名、jquery 选择器等；~{::selector} 表示在当前页寻找。

上面所列是 Thymeleaf 中几种常见表达式。

${...}用于表达变量，或可以用于表达复杂变量，例如：

```
<p>hello<span th:text="${name}">Jone</span>.</p>
```

以上示例展示一个简单的变量展示 ${name} 表达式，或者用点分隔变量 ${city.code} 等。

3.3.2 判断

表达式中的值可以与 >、<、>=、<=、== 和 != 符号进行比较。注意，如果是在 XML 中，则不应使用 < 和 > 符号，而用 < 和 > 替代。

另一种替代方法是使用运算符存在的文本别名：

- gt(>);
- lt(<);
- ge(>=);
- le(<=)，而不是(!)；
- eq(==);
- neq/ne(!=)。

下面示例如何使用 th:if：

```
<span th:if="${student.age lt 12}">
低年级学生
</span>
```

null 值的比较，如下示例：

```
<div th:if="${student.name} == null">
学生不存在
</div>
```

取反使用 th:unless：

```
<div th:unless="${ student }">
    <div>不存在的显示..</div>
```

```
</div>
```

Thymeleaf 支持三元表达式,如下示例:

```
<p th:class="${name}? 'even' : 'odd'">test</p>
```

还可通过 th:switch 进行分支判断,例如:

```
<span th:switch="${stu.age}">
    <p th:case="10">10 岁</p>
    <p th:case="11">11 岁</p>
    <p th:case="*">NO!</p>
</span>
```

3.3.3 循环

如果遍历 List 集合,则配合 th:each 即可快速完成遍历,如下代码所示:

```
<table>
    <tr th:each="stu : ${students}">
        <td th:text="${stu.name}">xm</td>
        <td th:text="${stu.age}">12</td>
        <td th:text="${stu.present}? #{true} : #{false}">false</td>
    </tr>
</table>
```

stu:${students} 表示对于 ${students} 中的每个元素,使用变量 stu 重复其中的每一个值,然后在循环中分别进行处理。

在迭代过程中,可以方便地跟踪当前所处状态,这在开发中非常有用,即使用状态变量,其包含在 th:each 中,包括:

- Index 属性:开始为 0 的迭代索引;
- count 属性:开始为 1 的迭代索引;
- size 属性:元素总量;
- current 属性:每次迭代的 iter 变量;
- even/odd 布尔属性:判断当前迭代是偶数还是奇数;
- first 布尔属性:判断当前迭代是否是第一个;
- last 布尔属性:判断当前迭代是否是最后一个。

下面是一个简单示例代码:

```
<table>
    <tr th:each="stu,iterStat : ${students}" th:class="${iterStat.odd}? 'odd'">
        <td th:text="${stu.name}">xm</td>
        <td th:text="${stu.age}">12</td>
        <td th:text="${stu.present}? #{true} : #{false}">false</td>
    </tr>
</table>
```

在以上示例中,显式定义了状态变量,即 iterStat,如果不指定状态名称,程序将隐式创

建以当前迭代名称作为开头、以 Stat 结尾的状态名称,比如下面示例:

```
<table>
    <tr th:each="stu : ${students}" th:class="${stuStat.odd}? 'odd'">
        <td th:text="${stu.name}">xm</td>
        <td th:text="${stu.age}">12</td>
        <td th:text="${stu.present}? #{true} : #{false}">false</td>
    </tr>
</table>
```

在迭代过程中,可以结合 th:if 进行条件判断,控制某些内容是否显示。例如,学生信息中包含有照片,则显示,否则不显示。

注意:以上循环的书写方式。在这里,有多种可被迭代遍历的对象,包括但不限于:

- java.util.Iterable;
- java.util.Enumeration;
- java.util.Iterator;
- java.util.Map;
- 数组。

3.3.4 属性修饰符

在上面接触了 th:class,其代表 HTML 5 中的 class 属性,Thymeleaf 提供了很多类似的属性修饰符,如表 3.1 所示。

表 3.1 属性修饰符

th:abbr	th:accept	th:accept-charset
th:accesskey	th:action	th:align
th:alt	th:archive	th:audio
th:autocomplete	th:axis	th:background
th:bgcolor	th:border	th:cellpadding
th:cellspacing	th:challenge	th:charset
th:cite	th:class	th:classid
th:codebase	th:codetype	th:cols
th:colspan	th:compact	th:content
th:contenteditable	th:contextmenu	th:data
th:datetime	th:dir	th:draggable
th:dropzone	th:enctype	th:for
th:form	th:formaction	th:formenctype
th:formmethod	th:formtarget	th:fragment
th:frame	th:frameborder	th:headers
th:height	th:high	th:href
th:hreflang	th:hspace	th:http-equiv
th:icon	th:id	th:inline
th:keytype	th:kind	th:label

限于篇幅,表 3.1 只列出了部分属性修饰符。

3.3.5 内嵌对象

为了模板更加易用,Thymeleaf 提供了一系列内置对象(内置于 Context 中)。包括:
- dates:java.util.Date 的功能方法类;
- calendars:类似♯dates,面向 java.util.Calendar;
- numbers:格式化数字的功能方法类;
- strings:字符串对象的功能类,包括 contains、startWiths、prepending/appending 等;
- objects:对 objects 的功能类操作;
- bools:对布尔值求值的功能方法;
- arrays:对数组的功能类方法;
- lists:对 lists 功能类方法;
- sets:用于处理 set 的方法;
- maps:用于处理 map 的方法;
- aggregates:用于在数组或集合上创建聚合的方法;
- ids:用于处理可能重复的 id 属性的实用程序方法。

使用方式示例如下:

```
${♯dates.format(date,'dd/MMM/yyyy HH:mm')}
${♯dates.setFormat(datesSet,'dd/MMM/yyyy HH:mm')}
${♯calendars.format(cal,'dd/MMM/yyyy HH:mm')}
${♯calendars.arrayFormat(calArray,'dd/MMM/yyyy HH:mm')}
${♯calendars.listFormat(calList,'dd/MMM/yyyy HH:mm')}
${♯numbers.formatInteger(num,3,'POINT')}
${♯numbers.setFormatInteger(numSet,3,'POINT')}
${♯strings.defaultString(text,default)}
${♯strings.setDefaultString(textSet,default)}
${♯bools.isFalse(cond)}
${♯bools.arrayIsFalse(condArray)}
${♯arrays.toStringArray(object)}
${♯lists.isEmpty(list)}
${♯sets.isEmpty(set)}
${♯maps.isEmpty(map)}
${♯aggregates.avg(array)}
${♯ids.next('someId')}
```

3.3.6 基本配置

Spring Boot 2 重点是简化配置,实际上,在 Spring Boot 2 工程中引入 Thymeleaf,便可以直接按其默认配置使用。包括:
- 模板存放位置:/src/main/resources/templates;
- 模板后缀:html;

- 模板编码：UTF-8；
- 模板样式：HTML5（支持 HTML、XML、TEXT、JAVASCRIPT）；
- 缓存模式：true。

以上默认设置基本能满足开发需要，但开发人员可根据需要修改其中一些配置项。例如缓存模式默认设置为 true 时，在编辑模板时，重启服务才能使模板生效，增加开发时，工作量会比较大。此时，最好将其设置为 false。

设置方法是在配置文件 application.properties 中进行设置，如下代码所示：

```
spring.thymeleaf.mode = HTML5
spring.thymeleaf.encoding = UTF-8
spring.thymeleaf.content-type = text/html
spring.thymeleaf.cache = false
spring.thymeleaf.prefix = classpath:/templates
```

设置完成并保存配置文件，重启服务，使设置生效。

3.4　Thymeleaf 示例

本节将结合 Thymeleaf 的基本语法和设置方法进行详细示例的展示。建立工程目录结构如图 3.3 所示。

更改配置文件 application.properties，增加如下内容：

```
server.applicatonName = 学生名单
spring.thymeleaf.cache = false
```

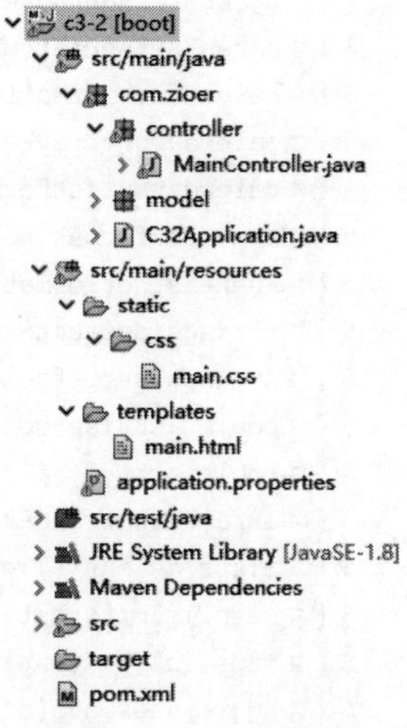

以上代码定义了示例名称和设置 thymeleaf 的缓存为 false，以防止修改 html 模板时，需重启服务。新增模型文件 Student.java，内容如下所示：

图 3.3　示例结构

```
package com.zioer.model;

import java.util.Date;

public class Student {
    private String name;
    private int age;
    private String grade;
    private Date birthday;

    public Student(String name, int age, String grade, Date birthday) {
        this.name = name;
        this.age = age;
        this.grade = grade;
        this.birthday = birthday;
    }

    //以下省略 get 和 set 方法
}
```

以上自定义简单学生 Model。新增控制文件 MainController,内容如下所示:

```
package com.zioer.controller;

//此处省略部分 import

import com.zioer.model.Student;

@Controller
public class MainController {

    @Value("${server.applicatonName}")
    String name;

    @GetMapping("main")
    public String main(Model model) throws ParseException {
        List<Student> list = new ArrayList<Student>();

        list.add(new Student("Jone", 11, "first", strtodate("2009-10-6")));
        list.add(new Student("Kitty", 12, "two", strtodate("2010-07-11")));
        list.add(new Student("Lucy", 10, "first", strtodate("2008-4-21")));
        list.add(new Student("Tom", 13, "three", strtodate("2010-12-4")));
        list.add(new Student("Jullu", 12, "two", strtodate("2007-5-13")));

        model.addAttribute("name",name);
        model.addAttribute("students", list);

        return "main";
    }

    private Date strtodate(String str) throws ParseException {
        SimpleDateFormat simpleDateFormat = new SimpleDateFormat("yyyy-MM-dd");
        return simpleDateFormat.parse(str);
    }
}
```

以上控制类 MainController 中,主要定义了 main 方法。为了简单说明,在此直接定义了 Student 实例,加入 List,并加入了自定义变量 model,包含属性 name 和 students,最后返回字符串 main,表示模板名称为 main.html;其中,自定义方法 strtodate 实现将字符串转化为日期。增加 CSS 文件 main.css,内容如下:

```
h2{
    text-align: center;
}
table
{
    border-collapse: collapse;
    margin: 0 auto;
    text-align: center;
}
```

```css
table td, table th
{
    border: 1px solid #cad9ea;
    color: #666;
    height: 30px;
}
table thead th
{
    background-color: #CCE8EB;
    width: 100px;
}
.even{
    background-color: #F5FAFA;
}
```

以上内容主要定义 html 文件显示内容样式。增加模板文件 main.html，内容如下：

```html
<!DOCTYPE html>
<html>
    <head>
        <meta charset="UTF-8" />
        <title>demo</title>
        <link rel="stylesheet" type="text/css" media="all" th:href="@{/css/main.css}" />
    </head>
    <body>
        <h2 th:text=""${name}">title</h2>
        <table style="width:80%">
            <thead>
            <tr>
                <th>序号</th>
                <th>姓名</th>
                <th>年龄</th>
                <th>所在年级</th>
                <th>出生年月</th>
            </tr>
            </thead>
            <tr th:each="stu,iterStat : ${students}" th:class="${iterStat.even}? 'even'">
                <span th:unless="${iterStat.last}">
                <td th:text="${iterStat.count}">xh</td>
                <td th:text="${stu.name}">xm</td>
                <td th:text="${stu.age}">12</td>
                <td th:text="${stu.grade}">年级</td>
                <td th:text="${#dates.format(stu.birthday, 'yyyy-MM-dd')}">birthday</td>
                </span>
                <span th:if="${iterStat.last}">
                    <td colspan="5" th:text="'合计：' + ${iterStat.size}">xm</td>
                </span>
            </tr>
        </table>
    </body>
</html>
```

以上模板文件实际上就是 html 文件，只是在某些元素中加入特殊属性。例如，包含了 Thymeleaf 中一些常用的属性修饰符，包括 th:href、th:text、th:each、th:class、th:unless 和 th:if 等。如果直接用浏览器打开该文件，显示如图 3.4 所示。

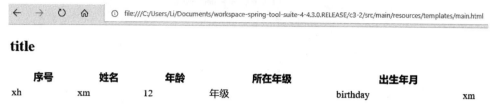

图 3.4　浏览器直接打开模板文件显示内容

如图所示，浏览器不会报错误，基本所见即所得。这实际是有利于页面设计师进行设计。保存以上文件，然后启动工程，在浏览器中访问如下网址：

http://127.0.0.1:8080/main

浏览器显示结果如图 3.5 所示。显示内容和预期结果相同。

序号	姓名	年龄	所在年级	出生年月
1	Jone	11	first	2009-10-06
2	Kitty	12	two	2010-07-11
3	lucy	10	first	2008-04-21
4	Tom	13	three	2010-12-04
合计：5				

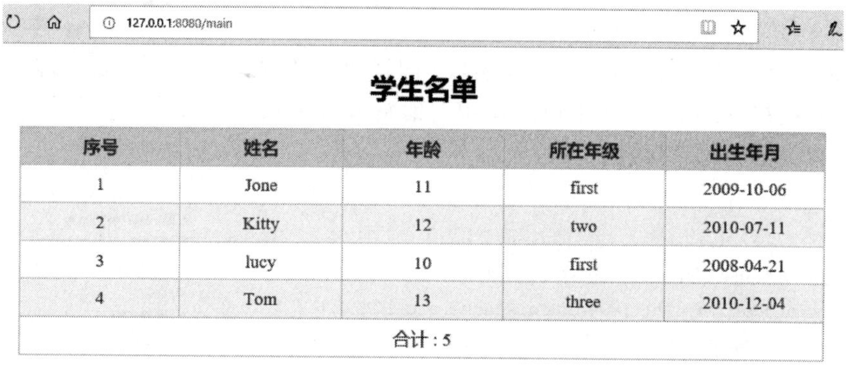

图 3.5　访问结果

3.5　体验 FreeMarker

FreeMarker 也是一款模板引擎，面向开发人员，其用来生成输出文本，包括 HTML 页面、源代码、电子邮件、配置文件、源代码等。它是一个 Java 类库，是一款开发人员可以嵌入到他们所开发产品的组件。

FreeMarker 模板默认后缀为 ftl(FreeMarker Template Language 的缩写)的文件。它是简单的、专用的语言，同样，它不关心数据的产生过程，只关心数据的显示。开发人员可专注于自己的领域。其工作原理如图 3.6 所示。

FreeMarker 产生时间较长，其第一个版本 1999 年末便出现，直到现在还在维护中。并已被业界很多项目所使用，作者也很喜爱该模板，在很多实际项目中使用 FreeMarker 做前端展示。

要在 Spring Boot 2 项目中使用 FreeMarker，需要在 pom.xml 文件中加入 FreeMarker

图 3.6　FreeMarker 工作原理

的依赖，如下代码所示：

```xml
<dependency>
    <groupId>org.springframework.boot</groupId>
    <artifactId>spring-boot-starter-freemarker</artifactId>
</dependency>
```

加入以上代码，Spring Boot 2 自动加入了能运行和使用 FreeMarker 的相关依赖，此时项目已能使用 FreeMarker 模板。本示例结构如图 3.7 所示。

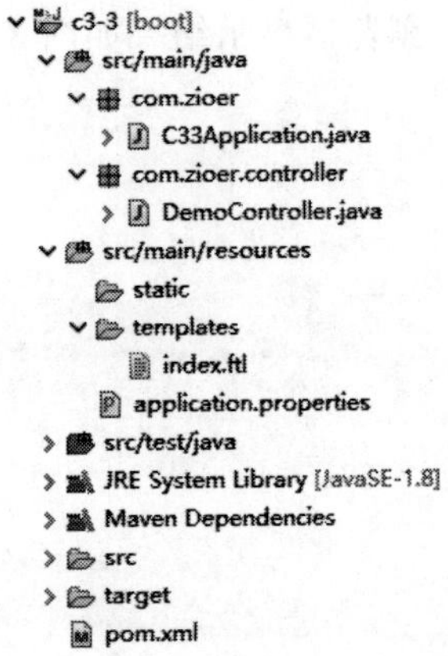

下面是几个主要文件示例。Freemarker 同样可以不做任何配置，直接在 Spring Boot 中快速使用。编辑 pom.xml 文件，加入相关依赖，如下代码所示：

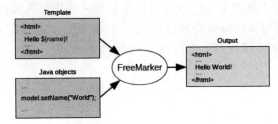

图 3.7　FreeMarker 示例结构

```xml
<project
xmlns="http://maven.apache.org/POM/4.0.0"
xmlns:xsi="http://www.w3.org/2001/XMLSchema-instance"
xsi:schemaLocation="http://maven.apache.org/POM/4.0.0
http://maven.apache.org/xsd/maven-4.0.0.xsd">
    <modelVersion>4.0.0</modelVersion>
    <parent>
        <groupId>org.springframework.boot</groupId>
        <artifactId>spring-boot-starter-parent</artifactId>
        <version>2.2.6.RELEASE</version>
        <relativePath/> <!-- lookup parent from repository -->
    </parent>
    <groupId>com.zioer</groupId>
    <artifactId>c3-3</artifactId>
    <version>0.0.1-SNAPSHOT</version>
    <packaging>jar</packaging>
    <name>c3-3</name>
    <description>demo</description>

    <properties>
        <java.version>1.8</java.version>
    </properties>

    <dependencies>
        <dependency>
            <groupId>org.springframework.boot</groupId>
```

```xml
            <artifactId>spring-boot-starter-freemarker</artifactId>
        </dependency>
        <dependency>
            <groupId>org.springframework.boot</groupId>
            <artifactId>spring-boot-starter-web</artifactId>
        </dependency>

        <dependency>
            <groupId>org.springframework.boot</groupId>
            <artifactId>spring-boot-starter-test</artifactId>
            <scope>test</scope>
        </dependency>
    </dependencies>

    <build>
        <plugins>
            <plugin>
                <groupId>org.springframework.boot</groupId>
                <artifactId>spring-boot-maven-plugin</artifactId>
            </plugin>
        </plugins>
    </build>
</project>
```

以上代码加入了 FreeMarker 等相关依赖。增加控制类文件 DemoController.java，代码如下所示：

```java
package com.zioer.controller;

import org.springframework.stereotype.Controller;
import org.springframework.ui.Model;
import org.springframework.web.bind.annotation.GetMapping;

@Controller
public class DemoController {

    @GetMapping("index")
    public String index(Model model) {
        model.addAttribute("title","freeMarker 示例");
        return "index";
    }
}
```

以上 Java 文件很简单，将该文件定义为控制类文件，即 DemoController 文件，在类名上加上注解@Controller，类中加入 index()方法，标记注解@GetMapping，该方法中加入 Model 模型，添加属性 title。最后返回 index 字符串，实际上，返回的是文件 index.ftl。

增加模型文件 index.ftl，默认放在文件夹 templates 中。编辑该文件，内容如下：

```
<!DOCTYPE html>
<html>
<head lang="en">
```

```
        < meta charset = "UTF - 8" />
        < title ></ title >
</ head >
< body >
    < h1 >$ {title}</ h1 >
</ body >
</ html >
```

以上代码内容很简单,FreeMarker 模板书写方式如 \${title} 所示。注意 \$ 书写方式。部署该项目,并在浏览器中访问,运行结果如图 3.8 所示。

图 3.8 freeMarker 示例

由此,FreeMarker 的使用不需要过多的设置,便能快速启动,查看效果。

3.6 FreeMarker 语法

本节介绍 FreeMarker 常用语法。

3.6.1 基本规则

模板运行的基本方式是插值,首先要了解如何进行插值,其语法是:

\${…}

正如前面示例,\${title} 表示将用变量 title 的实际值进行替换,\${…} 是 FreeMarker 最基本,也最常用的语法。

其次,注释在一门语言中也是很重要的部分,FreeMarker 的注释基本方式如下:

```
<# -- … -->
```

而 HTML 的注释方式是<!-- … -->。

以上两种注释的区别在于,FreeMarker 的注释在最终页面不输出,而 HTML 的注释将在页面直接输出,对于一些不希望最终页面输出的注释,建议使用 FreeMarker 注释方式。

基本规则中,同样重要的是 FTL(FreeMarker Template Language 缩写)指令标签,其基本方式如下:

```
<# ftl …>
…
</# ftl>
```

以上是 FTL 的基本形式，ftl 是各种指令名称，类似于 HTML 标签，其主要目的是用于行为控制，在页面不输出，区别于 HTML 标签，利于 FreeMarker 能识别这样的 FTL 指令。例如：

```
<#if condition>
    ...
<#elseif condition2>
    ...
<#else>
    ...
</#if>
```

以上是 FreeMarker 最基本的显示形式，当然还有其他的一些基本形式，后面涉及时将再进行介绍。

3.6.2　字符输出

字符输出是一种常见方式，形如：

${expr}

变量 expr 可以是单值，也可是对象中的值，例如：

```
${name}                    //输出变量 name 的值
${stu.name}                //输出对象 stu 中 name 的值
```

如果 expr 不存在，或值为 null 时，采用以上方式，页面会输出错误。为了避免这种情况，可采用下面方式：

${name?if_exists}

或

```
${name!}
${name?default("abc")}     //如果变量存在，输出该变量，否则不输出
```

或

${name!"xxx"}   //如果变量不存在，将输出指定默认值 abc

3.6.3　数字格式输出

数字输出方式和字符输出方式类似，形如：

${expr}

如果数字输出不进行处理或相关设置，则输出不一定符合要求，比如变量 sum 的值为 312322，则采用下面形式：

${sum}

其输出为：

312,322

以上输出格式不是任何时候都需要的。此时，需要对数字输出进行设置。
下面是数字输出的一些常用方式：

```
${sum?string.percent}              //输出按百分数显示
${sum?string.currency}             //输出按货币形式显示
${sum?string.number}               //输出带有千位分隔符
${sum?int}                         //转换为int型显示
${sum?string(",##0.0#")}           //自定义显示,表示小数点后最少1位,最多2位,小数点前
                                   //表示最少一位,含千位并以","分隔
```

设置页面默认显示方式，在页面顶部可增加如下设置：

```
<#setting number_format = "percent"/>       //设置输出方式为百分比
<#setting number_format = "#">              //输出数字时不显示千位分隔符
<#setting number_format = ",##0.##">        //自定义数字格式输出
```

如果一个页面中偶尔有数字需要格式化时，可以使用单个数字格式化，如果是一个页面都需要相同格式化，可以在页面设置默认的数字格式化形式。

3.6.4 日期格式输出

FreeMarker中日期型等价于Java中的Date类型，但不同之处在于不能直接输出，需要转换成字符串再输出。形式如下所示：

```
${date3?date}                                    //显示日期
${date3?time}                                    //显示时间
${date3?datetime}                                //显示日期时间复合
${date3?string('yyyy-MM-dd HH:mm:ss')}           //自定义形式
```

变量date3来自后端传递日期型，如果不转换为字符型输出，页面将报错。

3.6.5 其他数值

其他数值包括布尔值、集合等。
布尔值表达：true、false。
集合以方括号包含，各集合元素之间以英文逗号","分隔，例如：

```
<#assign a = ["one", "two", "three", "foue", "five", "six", "seven"]>
```

Map对象使用花括号包含，键和值之间以冒号":"分隔，键值对间以逗号","分隔，例如：

```
{"书":1000, "笔":800,"纸":300}
```

3.6.6 运算符

FreeMarker 运算符包括算数运算符、比较运算符和逻辑运算符。

算数运算符包括：

- 加：+；
- 减：-；
- 乘：*；
- 除：/；
- 取模：%。

例如：

```
<#assign a = 10>
${ a * a - 100 }
${ a /2 + a }
${ 13 % 10 }
```

比较运算符包括：

- =或者==：用于判断两个值是否相等；
- !=：用于判断两个值是否不等；
- >或者 gt：用于判断左边值是否大于右边值；
- >=或者 gte：用于判断左边值是否大于或等于右边值；
- <或者 lt：用于判断左边值是否小于右边值；
- <=或者 lte：用于判断左边值是否小于或等于右边值。

逻辑运算符包括：

- &&：逻辑与；
- ||：逻辑或；
- !：逻辑非。

逻辑运算符只作用于布尔值，否则会产生错误。

3.6.7 页面变量

数据可来自于后端，页面内也可声明和赋值变量，其语法如下：

```
<#assign … />
```

可以一次声明一个变量，或同时声明多个变量。例如，声明并赋值变量 age 的方式如下：

```
<#assign age = 30/>
${age}
```

页面变量可理解为局部变量，从后端传递的变量理解为全局变量。如果在页面声明的变量和全局变量相同，它们之间没有被覆盖，而是可以分别访问，示例如下所示：

```
<#assign title = "页面变量"/>    //后端传递已有title变量,此时页面创建相同变量
${title}                         //此时输出的将是页面局部变量
${.globals.title}                //输出的是全局变量的值
```

即使用.globals可访问全局变量。

3.6.8 判断指令

FreeMarker的判断指令包括if和switch。if基本形式是:

```
<#if condition>
  ...
<#elseif condition2>
  ...
<#else>
  ...
</#if>
```

代码示例:

```
<#assign x = 2/>
<#if x == 2>
  x 是 2
<#else>
  x 不是 2
</#if>
```

另一种方式是用switch方式进行选择判断,基本形式如下:

```
<#switch value>
  <#case refValue1>
    ...
    <#break>
  <#case refValue2>
    ...
    <#break>
  ...
  <#case refValueN>
    ...
    <#break>
  <#default>
    ...
</#switch>
```

示例代码如下所示:

```
<#assign y = 2/>
<#switch y>
  <#case 1>
    y = 1
    <#break>
```

```
<#case 2>
    y = 2
    <#break>
<#default>
    not
</#switch>
```

3.6.9　循环遍历

FreeMarker 循环遍历的基本形式如下:

```
<#list sequence as item>
    …
<#else>
    …
</#list>
```

代码示例:

```
<#assign a = ["one", "two", "three", "four", "five", "six", "seven"]>
<#list a as x>
    ${x}<br>
<#else>
    没值
</#list>
```

其中,<#else>可以不存在,只有没有迭代项时,该项用于输出特殊内容。

另一种形式如下:

```
<#list sequence>
    …
    <#items as item>
        …
    </#items>
    …
<#else>
    …
</#list>
```

以上形式可理解为,在循环体之前和之后有特殊显示,但当循环体内容为空时,不会只输出前后内容。示例代码如下:

```
<#list a>
  <ul>
    <#items as x>
      <li>${x}</li>
    </#items>
  </ul>
<#else>
    没值
</#list>
```

以上变量 x 为空时,将不会输出空 `<ul></ul>`。可以使用 break 指令在迭代的任意点退出,例如下面代码:

```
<#list 1..15 as x>
  ${x}
  <#if x == 4>
    <#break>
  </#if>
</#list>
```

break 对于运行中跳出循环很有用。在循环中,有时需要访问当前状态,包括当前迭代索引、判断是否最后一项等,此时,用到循环变量的内置方法,下面是示例代码:

```
<#list a as x>
  ${x?counter} : ${x}<br>
</#list>
```

上面代码中,内置方法 counter 表示当前索引从 1 开始。其他常用内置方法包括:
- has_next:表示是否是最后一项;
- index:表示索引,从 0 开始;
- is_even_item:循环项是否是当前迭代间隔 1 的奇数项;
- is_first:表示是否是第一项;
- is_last:表示是否是最后一项;
- is_odd_item:循环项是否是当前迭代间隔 1 的偶数项;
- item_parity:根据当前迭代间隔为 1 的索引的奇偶性,返回字符串值"odd"或"even"。

3.6.10 基本设置

Spring Boot 2 的特点是开箱即用,FreeMarker 一样,在工程中加入该依赖后,不需要任何设置,便可以直接使用,一般采用默认设置即可。例如,模板存放位置为:

/src/main/resources/templates

模板后缀为 ftl

FreeMarker 也提供了灵活设置,开发人员可以对这些常用设置进行更改,以更加符合习惯,设置放在项目配置文件 application.properties 中。常用设置如下:

```
spring.freemarker.allow-request-override=false  # 设置是否允许 HttpServletRequest 属性覆
#盖(隐藏)控制器生成的同名模型属性
spring.freemarker.allow-session-override=false  # 设置是否允许 HttpSession 属性覆盖(隐
#藏)控制器生成的同名模型属性
spring.freemarker.cache=false                    # 设置是否启用模板缓存
spring.freemarker.charset=UTF-8                  # 模板编码
spring.freemarker.check-template-location=true   # 检查模板位置是否存在
spring.freemarker.content-type=text/html         # Content-Type 值
spring.freemarker.enabled=true                   # 设置是否允许 MVC 使用
spring.freemarker.expose-request-attributes=false # 设置是否应在与模板合并之前将所有
```

请求属性添加到模型中
spring.freemarker.expose-session-attributes=false # 设置是否应在与模板合并之前将所有
HttpSession 属性添加到模型中
spring.freemarker.expose-spring-macro-helpers=true # 设置是否以 springMacroRequestContext
的形式暴露 RequestContext 给 Spring's macro library 使用
spring.freemarker.prefer-file-system-access=true # 首选文件系统访问模板加载.文件系统
# 访问可以热检测模板更改.
spring.freemarker.prefix= # 在构建 URL 时添加前缀
spring.freemarker.request-context-attribute= # 所有视图的 RequestContext 属性的名称
spring.freemarker.settings.*= # 设置 FreeMarker keys
spring.freemarker.suffix= # 在构建 URL 时附加到视图名称的后缀
spring.freemarker.template-loader-path=classpath:/templates/ # 设置模板的加载路径,多
个路径间以逗号分隔,默认:["classpath:/templates/"]
spring.freemarker.view-names= # 指定使用模板的视图列表

例如,统一设置数字显示形式,在配置文件 application.properties 中设置示例如下:

spring.freemarker.settings.number_format=0.##

限于篇幅,以上只介绍了 FreeMarker 作为模板语言的常用语法和设置。

3.7 FreeMarker 示例

本节将介绍一个使用 FreeMarker 模板的完整示例,结构如图 3.9 所示。

图 3.9 示例完整结构

编辑 pom.xml 文件,其完整内容如下所示:

```
<project
   xmlns="http://maven.apache.org/POM/4.0.0"
   xmlns:xsi="http://www.w3.org/2001/XMLSchema-instance"
```

```xml
xsi:schemaLocation = "http://maven.apache.org/POM/4.0.0
http://maven.apache.org/xsd/maven-4.0.0.xsd">
    <modelVersion>4.0.0</modelVersion>
    <parent>
        <groupId>org.springframework.boot</groupId>
        <artifactId>spring-boot-starter-parent</artifactId>
        <version>2.2.6.RELEASE</version>
        <relativePath/> <!-- lookup parent from repository -->
    </parent>
    <groupId>com.zioer</groupId>
    <artifactId>c3-4</artifactId>
    <version>0.0.1-SNAPSHOT</version>
    <packaging>jar</packaging>
    <name>c3-4</name>
    <description>demo</description>

    <properties>
        <java.version>1.8</java.version>
    </properties>

    <dependencies>
        <dependency>
            <groupId>org.springframework.boot</groupId>
            <artifactId>spring-boot-starter-freemarker</artifactId>
        </dependency>
        <dependency>
            <groupId>org.springframework.boot</groupId>
            <artifactId>spring-boot-starter-web</artifactId>
        </dependency>
        <dependency>
            <groupId>org.springframework.boot</groupId>
            <artifactId>spring-boot-starter-test</artifactId>
            <scope>test</scope>
        </dependency>
    </dependencies>

    <build>
        <plugins>
            <plugin>
                <groupId>org.springframework.boot</groupId>
                <artifactId>spring-boot-maven-plugin</artifactId>
            </plugin>
        </plugins>
    </build>
</project>
```

在以上代码中，重点是加入了 spring-boot-starter-freemarker 依赖。编辑配置文件 application.properties，加入下面内容：

```
server.applicationName = 学生详细信息                    //设置自定义项目名称
spring.freemarkerSettings.number_format = 0.##         //全局设置 FreeMarker 数字显示
```

根据需要,可以加入其他相关设置。建立模型文件 Student.java,内容如下所示:

```java
public class Student {
    private String name;
    private int age;
    private String grade;
    private Date birthday;
    private double regmoney;

    public Student(String name, int age, String grade, Date birthday, double regmoney) {
        this.name = name;
        this.age = age;
        this.grade = grade;
        this.birthday = birthday;
        this.regmoney = regmoney;
    }
    //省略 get 和 set 方法
}
```

上面示例模型 Student 很容易理解,包括字符型、数字型、日期型和小数型。建立控制类文件 MainController.java,内容如下:

```java
@Controller
public class MainController {
    @Value("${server.applicationName}")
    String name;

    @GetMapping("main")
    public String main(Model model) throws ParseException {
        List<Student> list = new ArrayList<Student>();
        Date now = new Date();

        list.add(new Student("小小", 13, "初三", strtodate("2006-10-6"), 1003.6));
        list.add(new Student("天天", 12, "初一", strtodate("2007-7-11"), 1206.34));
        list.add(new Student("明明", 10, "初一", strtodate("2009-4-21"), 1134.6));
        list.add(new Student("淼淼", 13, "初三", strtodate("2006-12-4"), 1028.3));
        list.add(new Student("露露", 12, "初二", strtodate("2007-9-13"), 1323.4));
        list.add(new Student("芸芸", 11, "初一", strtodate("2008-5-13"), 1532.1));

        model.addAttribute("name", name);
        model.addAttribute("now", now);
        model.addAttribute("students", list);

        return "main";
    }

    private Date strtodate(String str) throws ParseException {
        SimpleDateFormat simpleDateFormat = new SimpleDateFormat("yyyy-MM-dd");
        return simpleDateFormat.parse(str);
    }
}
```

上面代码中加入了当前日期,同时,在业务模型中加入了不同类型数值,传递给前端,返回字符串main。main.ftl内容如下:

```html
<!DOCTYPE html>
<html>
    <head>
        <meta charset="UTF-8" />
        <title>demo</title>
        <link rel="stylesheet" type="text/css" media="all" href="/css/main.css" />
        <style type="text/css">

        </style>
    </head>
    <body>
        <h3>${name}</h3>
        <p></p>
        <table class="table-1">
            <caption>
            制表日期:${now?datetime}
            </caption>
            <thead>
            <tr>
                <th>序号</th>
                <th>姓名</th>
                <th>年龄</th>
                <th>所在年级</th>
                <th>出生年月</th>
                <th>学费</th>
            </tr>
            </thead>
            <#list students as student>
            <tr class="${student?item_parity}Row">
                <td>${student?counter}</td>
                <td>${student.name}</td>
                <td>${student.age}</td>
                <td>${student.grade}</td>
                <td>${student.birthday?date}</td>
                <td>${student.regmoney?string.currency}</td>
            </tr>
            <#if (student?is_last == true)>
            <tr>
                <td colspan="6">合计:${students?size}条</td>
            <tr>
            </#if>
            </#list>
        </table>
    </body>
</html>
```

上面代码比较典型,包括日期、数字的显示,处理循环的方式,循环内变量的利用等。下面是样式文件main.css:

```css
body {
    text-align: center;
```

```css
}
table.table-1 {
    font-family: verdana,arial,sans-serif;
    font-size:11px;
    color:#333333;
    border-width: 1px;
    border-color: #999999;
    border-collapse: collapse;
    width: 80%;
    margin: auto;
}
table.table-1 caption{
    text-align: right;
}
table.table-1 th {
    background:#b5cfd2;
    border-width: 1px;
    padding: 8px;
    border-style: solid;
    border-color: #999999;
}
table.table-1 td {
    border-width: 1px;
    padding: 8px;
    border-style: solid;
    border-color: #999999;
}
.evenRow{
    background:#eeeeee;
}
```

以上工程建立完成并运行后,浏览器输入:

http://127.0.0.1:8080/main

显示内容如图3.10所示。

序号	姓名	年龄	所在年级	出生年月	学费
1	小小	13	初三	2006-10-6	¥1,003.60
2	天天	12	初一	2007-7-11	¥1,206.34
3	明明	10	初一	2009-4-21	¥1,134.60
4	淼淼	13	初三	2006-12-4	¥1,028.30
5	露露	12	初二	2007-9-13	¥1,323.40
6	芸芸	11	初一	2008-5-13	¥1,532.10
合计: 6条					

学生详细信息
制表日期: 2019-7-17 21:41:57

图3.10 FreeMarker示例运行图

以上展示了 FreeMarker 模板完整示例，涉及的知识点比较多，也较典型，但由此可见，FreeMarker 在发展过程中，其体系已很完善。结合上面示例，可很快掌握 FreeMarker 的使用。

3.8 JSP 介绍

JSP 全名为 Java Server Pages，其本质是 servlet，即 JSP 拥有 servlet 的所有功能。它是在传统的网页 HTML 文件（*.htm，*.html）中插入 Java 程序段（Scriptlet）和 JSP 标签（tag），以形成 JSP 文件，后缀名为（*.jsp），实现业务层和视图层分离，这样，JSP 只负责显示数据即可，修改业务代码不会影响 JSP 页面代码。

首次通过浏览器访问 JSP 页面时，服务器会对该页面中代码进行编译，并且仅执行一次编译，编译后被保存，下次访问时直接执行编译过的代码，能节约服务器资源，提升客户端访问速度。

JSP 页面具有以下优点：
- 可以写 Java 代码；
- 支持 JSP 标签（jsp tag）；
- 支持表达式语言（el）；
- 官方标准。用户群广，可自定义便签库，以及丰富的第三方 JSP 标签库；
- 性能良好。JSP 编译成 Class 文件执行，有很好的性能表现。

缺点有：
- 前端开发不所见即所得；
- 需要编译。

在当前互联网高速发展中，前后分离越来越明显，特别是模板技术出现，模板技术可理解为一个占位符动态替换技术。JSP 出现较早，尽管体现了前后分离思想，但不够彻底，还需要动态编译，页面中能嵌入 Java 代码，团队协作开发复杂。在 Spring Boot 2 中，官方已不建议使用 JSP 技术，但不是说 Spring Boot 2 不支持该技术，特别是在 MVC 模式演进过程中，JSP 起到很重要作用。同时，很多开发人员还在使用 JSP，以及有很多业务系统还是采用 JSP 技术。作者希望，开发人员多接触该技术以及各种技术以适合不同业务场景，比如，中小规模系统采用 JSP 也应为一个好的选择。

但不管怎样，JSP 技术是一种页面处理的优秀的解决方案。

3.9 JSP 语法

JSP 也是嵌入到 HTML 页面中的脚本程序，其可以包含 Java 语句、变量、方法或表达式，语法格式如下：

```
<% … %>
```

或采用 XML 形式：

```
<jsp:scriptlet>
    ...
</jsp:scriptlet>
```

但一般简洁写法采用第一种形式。非脚本以外的文本，包括 HTML、JS 等，写在该脚本代码的外部。例如，简单的打印输出代码示例：

```
<%
    out.println("hello zioer");
%>
```

就这么简单。

JSP 注释的语法格式如下：

```
<%-- ... --%>
```

同样，JSP 注释不在最终 HTML 输出页面显示。JSP 提供对 Java 代码的全面支持，即 Java 代码可以在 JSP 页面中使用和运行，语法格式和 Java 相同。

声明方法如下：

```
<%! ... %>
```

或

```
<jsp:declaration>
    ...
</jsp:declaration>
```

比如：

```
<%! int k = 1; %>
```

JSP 中判断语句使用 If...else 块。

```
<%
int k = 1;
if(k==1){
    out.println("k = 1");
}else{
    out.println("k <> 1");
}
%>
```

熟悉 Java 代码，很容易理解上面代码。在 JSP 脚本中，循环语句使用 Java 里的三种基本循环类型：for、while 和 do…while。

在这里不再分别举例，只需要将相关语句放入脚本程序中，以使得 JSP 引擎能识别出可编译语句。

JSP 中重要的是提供了标准标签库(JSTL)，即 JSP 标签集合，封装了 JSP 应用的通用核心功能。同时，第三方也提供了标签库，便于程序的快速开发。

3.9.1 核心标签

核心标签是常用的 JSTL 标签。要使用核心标签,首先在页面顶部书写如下声明代码以引用该库:

```
<%@ taglib prefix="c" uri="http://java.sun.com/jsp/jstl/core" %>
```

该库中包含以下常用方法:
<c:out>标签:用于输出一个表达式结果;
<c:set>标签:用于保存数据;
<c:remove>标签:用于删除数据;
<c:if>标签:判断标签;
<c:choose>、<c:when>和<c:otherwise>标签:条件判断;
<c:import>标签:引入 URL 内容到当前页面;
<c:forEach>标签:迭代标签;
<c:redirect>标签:用于重定向到一个新的 URL.;
<c:url>标签:用于创造一个 URL。

使用方法示例如下:

```
<c:out value="&lt 字符串 &gt" escapeXml="true" default="默认值"></c:out>
<c:url value="/css/main.css" var="jstlCss" />
<link href="${jstlCss}" rel="stylesheet" />
<c:forEach var="i" begin="1" end="10">
    <c:out value="${i mod 2}"/><br>
</c:forEach>
```

3.9.2 格式化标签

该标签用来格式化文本、日期、时间、数字并输出到页面。需要在页面顶部书写如下声明代码引用该库:

```
<%@ taglib prefix="fmt" uri="http://java.sun.com/jsp/jstl/fmt" %>
```

该库中包含以下常用方法:
<fmt:formatNumber>标签:用于格式化数字,使用指定的格式或精度;
<fmt:formatDate>标签:用于格式化日期和时间,使用指定的风格;
<fmt:parseNumber>标签:用于解析一个数字、货币或百分比的字符串;
<fmt:parseDate>标签:用于解析日期或时间的字符串;
<fmt:setLocale>标签:用于指定地区;
<fmt:message>标签:用于显示资源配置文件信息。

使用方法如以下代码所示:

```
<c:set var="num" value="35674.328" />
```

```
<p><fmt:formatNumber value="${num}" type="currency"/></p>
<c:set var="now" value="<%=new java.util.Date()%>"/>
<p><fmt:formatDate type="both" value="${now}"/></p>
<fmt:parseNumber var="m" type="number" value="${num}"/>
<p><c:out value="${m}"/></p>
<fmt:parseDate value="${dt}" var="k" pattern="yyyy-MM-dd"/>
<p>解析后的日期为：<c:out value="${k}"/></p>
```

3.9.3 JSTL 函数

JSP 中包含了标准函数,大部分是通用的字符串处理函数,方便开发时快速调用和处理。同样,需要使用这些标准函数时,需要在页面顶部加入下面内容：

```
<%@ taglib prefix="fn" uri="http://java.sun.com/jsp/jstl/functions" %>
```

该库中包含以下常用方法：

fn:trim()：用于删除指定字符前后的空白符；
fn:startsWith()：用于检测字符串是否以指定的字符串开始；
fn:endsWith()：用于检测字符串是否以指定的字符串结尾；
fn:toLowerCase()：用于将字符串转为小写；
fn:toUpperCase()：用于将字符串转为大写；
fn:substring()：返回字符串子集；
fn:contains()：用于检测字符串是否包含指定的子串；
fn:indexOf()：用于返回指定字符串在源字符串中出现的位置；
fn:join()：用于合成字符串；
fn:length()：用于返回字符串长度；
fn:replace()：用于替换字符串；
fn:split()：用于分隔字符串并返回数组。

使用示例如下：

```
<c:set var="str1" value=" Hello Zioer "/>
<p> trim str1 后 长度 : ${fn:length( fn:trim(str1) )}</p>
<c:set var="str2" value="Hello Zioer."/>
<c:if test="${fn:startsWith(str2, 'Hello')}">
    <p>字符串 str2 以 Hello 开头</p>
</c:if>
<p> Index str1 : ${fn:indexOf(str1, "Hello")}</p>
<p>小写后: ${fn:toLowerCase(str1)}</p>
```

在 JSTL 中,还提供 SQL 标签,主要用于和关系型数据库交互,比如 MySQL 等；XML 标签,用于创建和操作 XML 文档。

3.9.4 Spring 标签库

Spring 标签库属于第三方标签库。其主要提供了两个标签库,一个是通用工具标签,

在页面使用时,需要在页面顶部书写下面声明代码:

```
<%@ taglib uri="http://www.springframework.org/tags" prefix="spring" %>
```

另一个是表单标签库,使用时,需要在页面顶部书写下面声明代码:

```
<%@ taglib uri="http://www.springframework.org/tags/form" prefix="form" %>
```

下面是 Spring 标签相关介绍。

spring:url 标签主要用于创建 URL,并将其赋值给一个变量,下面是示例代码:

```
<spring:url value="/css/main.css" var="springCss" />
<link href="${springCss}" rel="stylesheet" />
```

spring:escapeBody 标签主要用于转义内容,是一个通用的转义标签,支持 HTML 和 JavaScript 转义,下面是示例代码:

```
<spring:escapeBody htmlEscape="true">
    <h1>Hello zioer</h1>
</spring:escapeBody>
<spring:escapeBody javaScriptEscape="true">
    <script type="text/javascript">alert('hello zioer')</script>
</spring:escapeBody>
```

Spring 标签库中的 form 相关标签主要用于数据绑定,以减少开发工作量。

form:form 标签:开发人员指定该 form 绑定的是哪个 Model,例如:

```
<form:form modelAttribute="person" …>
```

以上示例中,属性 modelAttribute 指明该 form 绑定了哪一个 Model。只有 form 先绑定 Model,然后在该 form 标签内的其他表单标签中,才可以通过属性 path 指定 Model 属性的名称来绑定相关数据。

form:input 标签:生成一个 type 为 text 的 input 标签。例如:

```
<form:input path="name" />
```

以上示例中,属性 path 将绑定 form 指定模型中 name 属性的值。生成 Html 如下:

```
<input id="name" name="name" type="text" value="" />
```

如果该模型的 name 有值,则 value 中也会出现该值,这对于数据的绑定提供了便利。

form:password 标签:生成一个 type 为 password 的 input 标签,例如:

```
<form:password path="password" />
```

form:hidden 标签:生成一个隐藏的 input 标签,例如:

```
<form:hidden path="id" />
```

form:textarea 标签:生成多行输入的标签,例如:

```
<form:textarea path="notes" />
```

form:checkbox 标签:生成 checkbox 的标签,例如:

```
<form:checkbox path = "ckb"/>
```

form:checkboxes 标签：是一个选项组，生成多个复选框，例如：

```
<form:checkboxes items = "${favitems}" path = "favitem" />
```

其中，favitems 是从后端传入的 list、map 等类型，favitem 是模型中对应属性。

form:radiobutton 标签：生成 radio 的标签，例如：

```
<form:radiobutton path = "sex" value = "M"/>
```

form:radiobuttons 标签：是一个选项组，其生成多个单选按钮。例如：

```
<form:radiobuttons items = "${bookItem}" path = "books"/>
```

form:select 标签：用于生成 select 的标签，例如：

```
<form:select path = "country" items = "${countryItems}" />
```

form:option 标签：用于生成 select 中 option 的标签。
form:options 标签：用于生成 select 中一组 option 的标签，例如：

```
<form:select path = "book">
    <form:option value = " - " label = "—请选择—"/>
    <form:options items = "${books}" />
</form:select>
```

在上面示例中，可知 form:select 标签生成 select 的 option 项有两种方式。

form:errors 标签：用于生成一个或者多个 span 元素，每个 span 元素包含一个错误消息。例如：

```
<form:errors path = " * " cssClass = "errorbox" element = "div" />
<form:errors path = "name" cssClass = "error" element = "div"/>
```

以上介绍了 JSP 中基本语法以及标准标签、第三方标签语法及相关示例。通过介绍，JSP 提供了有助于开发人员快速开发业务系统的标签库，特别是其提供并允许第三方标签接入，比如 Spring 提供的标签组。

3.10 JSP 示例

在本节中，将结合上面介绍的知识，完成一个示例，以帮助理解上节所介绍知识点。建立一个基于 Maven 的 Java 项目，项目结构如图 3.11 所示。

由于该示例采用 JSP，在图 3.11 中，故增加了文件夹 webapp，类似于典型的 Java Web 项目结构。编辑项目中文件 pom.xml，内容如下所示：

```
<?xml version = "1.0" encoding = "UTF - 8"?>
<project
xmlns = "http://maven.apache.org/POM/4.0.0"
xmlns:xsi = "http://www.w3.org/2001/XMLSchema - instance"
```

```xml
xsi:schemaLocation = "http://maven.apache.org/POM/4.0.0
http://maven.apache.org/xsd/maven-4.0.0.xsd">
    <modelVersion>4.0.0</modelVersion>
    <parent>
        <groupId>org.springframework.boot</groupId>
        <artifactId>spring-boot-starter-parent</artifactId>
        <version>2.2.6.RELEASE</version>
        <relativePath/> <!-- lookup parent from repository -->
    </parent>
    <groupId>com.zioer</groupId>
    <artifactId>c3-5</artifactId>
    <version>0.0.1-SNAPSHOT</version>
    <packaging>war</packaging>
    <name>c3-5</name>
    <description>demo</description>

    <properties>
        <java.version>1.8</java.version>
    </properties>

    <dependencies>
        <dependency>
            <groupId>org.springframework.boot</groupId>
            <artifactId>spring-boot-starter-web</artifactId>
        </dependency>
        <dependency>
            <groupId>org.springframework.boot</groupId>
            <artifactId>spring-boot-starter-test</artifactId>
            <scope>test</scope>
        </dependency>
        <!-- Web with Tomcat + Embed -->
        <dependency>
            <groupId>org.springframework.boot</groupId>
            <artifactId>spring-boot-starter-tomcat</artifactId>
            <scope>provided</scope>
        </dependency>
        <!-- JSTL -->
        <dependency>
            <groupId>javax.servlet</groupId>
            <artifactId>jstl</artifactId>
        </dependency>

        <!-- Need this to compile JSP -->
        <dependency>
            <groupId>org.apache.tomcat.embed</groupId>
            <artifactId>tomcat-embed-jasper</artifactId>
            <!-- <scope>provided</scope> -->
        </dependency>
```

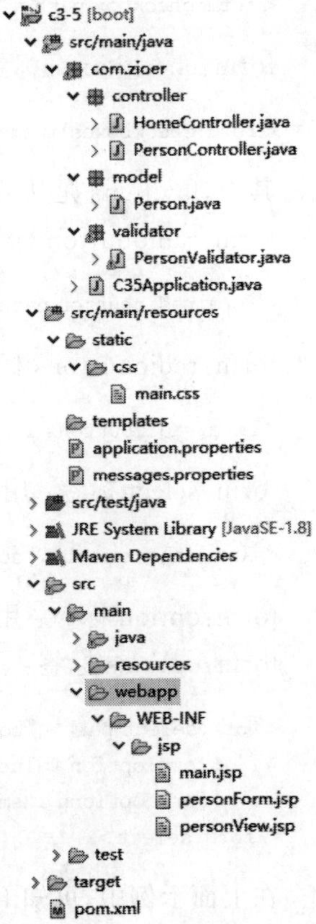

图 3.11 示例结构

```xml
<!-- Need this to compile JSP -->
<dependency>
    <groupId>org.eclipse.jdt.core.compiler</groupId>
    <artifactId>ecj</artifactId>
    <version>4.6.1</version>
    <scope>provided</scope>
</dependency>
<!-- Optional, for bootstrap -->
<dependency>
    <groupId>org.webjars</groupId>
    <artifactId>bootstrap</artifactId>
    <version>4.3.1</version>
</dependency>
<dependency>
    <groupId>org.webjars</groupId>
    <artifactId>jquery</artifactId>
    <version>3.4.1</version>
</dependency>
    </dependencies>
    <build>
        <plugins>
            <plugin>
                <groupId>org.springframework.boot</groupId>
                <artifactId>spring-boot-maven-plugin</artifactId>
            </plugin>
        </plugins>
    </build>

</project>
```

在以上代码中，加入了必需的 JSTL 相关依赖、前端相关依赖，以保证项目中能使用 JSTL，同时需要注意的是 packaging 方式是 war，而不要填写为 jar，否则，会造成最终生成 Jar 包无法找到 jsp 文件的问题。编辑配置文件 application.properties，增加如下内容：

```
spring.mvc.view.prefix: /WEB-INF/jsp/
spring.mvc.view.suffix: .jsp

#2.0 以后设置,编辑 jsp 文件后,自动编译
server.servlet.jsp.init-parameters.development=true
#2.0 以前设置
# server.jsp-servlet.init-parameters.development=true
welcome.message: 测试
```

提示：设置 server.servlet.jsp.init-parameters.development=true，用于保证编辑 jsp 文件后，不需要重启服务，能及时看到效果。

编辑控制类文件 HomeController.java，如下所示：

```
@Controller
public class HomeController {
    @Value("${welcome.message:test}")
```

```
    private String message = "Hello World";

    @GetMapping("/home")
    public String main(Model model){
        model.addAttribute("message", message);
        return "main";
    }
}
```

上面代码比较简单，读取配置文件 application.properties 中的值，并传递到前端页面，编辑前端页面 main.jsp，内容如下：

```
<!DOCTYPE html>
<%@ taglib prefix="c" uri="http://java.sun.com/jsp/jstl/core" %>
<%@ taglib prefix="fmt" uri="http://java.sun.com/jsp/jstl/fmt" %>
<%@ taglib prefix="fn" uri="http://java.sun.com/jsp/jstl/functions" %>

<html lang="en">
<head>
    <meta charset="UTF-8"/>
    <link rel="stylesheet" type="text/css"
        href="/webjars/bootstrap/4.3.1/css/bootstrap.min.css"/>
    <c:url value="/css/main.css" var="jstlCss"/>
    <link href="${jstlCss}" rel="stylesheet"/>
</head>

<body>
    <nav class="navbar navbar-expand-md navbar-dark bg-dark">
        <a class="navbar-brand" href="#">导航栏</a>
        <button class="navbar-toggler" type="button" data-toggle="collapse" data-target="#navbarsExampleDefault" aria-controls="navbarsExampleDefault" aria-expanded="false" aria-label="Toggle navigation">
            <span class="navbar-toggler-icon"></span>
        </button>

        <div class="collapse navbar-collapse" id="navbarsExampleDefault">
            <ul class="navbar-nav mr-auto">
                <li class="nav-item active">
                    <a class="nav-link" href="/home">首页</a>
                </li>
                <li class="nav-item">
                    <a class="nav-link" href="/person">增加人员</a>
                </li>
                <li class="nav-item">
                    <a class="nav-link disabled" href="#" tabindex="-1" aria-disabled="true">查看</a>
                </li>
            </ul>
            <form class="form-inline my-2 my-lg-0">
                <input class="form-control mr-sm-2" type="text" placeholder="关键字" aria-label="Search"/>
```

```html
            <button class="btn btn-secondary my-2 my-sm-0" type="submit">检索</button>
        </form>
    </div>
</nav>
<main role="main" class="container">
    <div class="starter-template">
        <h1>Spring Boot Web JSP Example</h1>
        <h2>Message: ${message}</h2>
        <c:out value="${message}"></c:out>
        <c:out value="&lt要显示的数据对象(未使用转义字符)&gt" escapeXml="true" default="默认值"></c:out><br/>

        <%
        int k = 1;
            if(k == 1){
                out.println("k = 1");
            }else{
                out.println("k <> 1");
            }
        %>
        <br>
        <c:forEach var="i" begin="1" end="10">
            <c:out value="${i mod 2}"/><br>
        </c:forEach>

        <c:set var="num" value="35674.328" />
            <p><fmt:formatNumber value="${num}"
            type="currency"/></p>
        <c:set var="now" value="<%= new java.util.Date() %>" />
        <p><fmt:formatDate type="both"
            value="${now}" /></p>

        <fmt:parseNumber var="i" type="number" value="${num}" />
        <p><c:out value="${i}" /></p>
        <c:set var="dt" value="2019-10-1" />
        <fmt:parseDate value="${dt}" var="k" pattern="yyyy-MM-dd" />
        <p>解析后的日期为: <c:out value="${k}" /></p>

        <c:set var="str1" value="    Hello Zioer            "/>
        <p>trim str1 后 长度: ${fn:length( fn:trim(str1) )}</p>
        <c:set var="str2" value="Hello Zioer."/>
        <c:if test="${fn:startsWith(str2, 'Hello')}">
            <p>字符串 str2 以 Hello 开头</p>
        </c:if>
        <p>Index str1: ${fn:indexOf(str1, "Hello")}</p>
        <p>小写后: ${fn:toLowerCase(str1)}</p>

    </div>
</main>
<script type="text/javascript"
    src="webjars/jquery/3.4.1/jquery.min.js"></script>
```

```
    < script type = "text/javascript"
        src = "webjars/bootstrap/4.3.1/js/bootstrap.min.js"></script>
</body>
</html>
```

上面代码没有成体系,主要是结合前面介绍JSTL相关内容进行展示,关键是该页面顶部加入相关库代码很重要,同时演示了如何加入Jar包中的bootstrap.min.css等文件。效果如图3.12所示。

图 3.12　访问 home 页面

编辑模型类 Person.java,主要用于演示 Spring 标签库,代码如下所示:

```
public class Person {

    private long id;

    private String name;
    private String email;
    private String dateOfBirth;

    @NotEmpty
    private String password;
    private String sex;
    private String regin;
    private String book;
    private String job;
    private boolean receiveNewsletter;
    private String[] hobbies;
    private List < String > favouriteLanguage;
    private List < String > fruit;
    private String notes;
    private MultipartFile file;

    //以下省略 get 和 set
}
```

编辑验证文件 PersonValidator.jsp,代码如下所示:

```java
@Component
public class PersonValidator implements Validator {

    @Override
    public boolean supports(final Class calzz) {
        return Person.class.isAssignableFrom(calzz);
    }

    @Override
    public void validate(final Object obj, final Errors errors) {

        ValidationUtils.rejectIfEmptyOrWhitespace(errors, "name", "required.name");
    }
}
```

编辑配置文件 messages.properties, 代码如下所示:

```
required.name = 姓名必须输入!
NotEmpty.person.password = 密码必须输入!
```

在这里, 仅测试了两个 input 不能为空的情况。编辑控制类文件 PersonController.java, 代码如下所示:

```java
@Controller
public class PersonController {
    @Autowired
    PersonValidator validator;

    @GetMapping("/person")
    public ModelAndView showForm(final Model model) {
        initData(model);
        return new ModelAndView("personForm", "person", new Person());
    }

    @PostMapping("/addPerson")
    public String submit(@Valid @ModelAttribute("person") final Person person, final BindingResult result, final ModelMap modelMap, final Model model) {
        validator.validate(person, result);
        if (result.hasErrors()) {
            initData(model);
            return "personForm";
        }

        modelMap.addAttribute("person", person);
        return "personView";
    }

    private void initData(final Model model) {

        final List<String> favouriteLanguageItem = new ArrayList<String>();
        favouriteLanguageItem.add("Java");
```

```java
            favouriteLanguageItem.add("C++");
            favouriteLanguageItem.add("Python");
            model.addAttribute("favouriteLanguageItem", favouriteLanguageItem);

            final List<String> jobItem = new ArrayList<String>();
            jobItem.add("全职");
            jobItem.add("兼职");
            model.addAttribute("jobItem", jobItem);

            final Map<String, String> reginItems = new LinkedHashMap<String, String>();
            reginItems.put("bj", "北京");
            reginItems.put("sh", "上海");
            reginItems.put("gz", "广州");
            reginItems.put("cd", "成都");
            model.addAttribute("reginItems", reginItems);

            final List<String> fruit = new ArrayList<String>();
            fruit.add("苹果");
            fruit.add("梨子");
            fruit.add("葡萄");
            model.addAttribute("fruit", fruit);

            final List<String> books = new ArrayList<String>();
            books.add("零基础C++学习");
            books.add("21天学会Python");
            books.add("红楼梦");
            model.addAttribute("books", books);
    }
}
```

以上代码初始化页面需要基础数据,创建两个方法,一个用于新增用户,一个用于展示新增用户数据。编辑jsp页面personForm.jsp,代码如下所示:

```jsp
<!DOCTYPE html>
<%@ taglib prefix="spring" uri="http://www.springframework.org/tags" %>
<%@ taglib prefix="form" uri="http://www.springframework.org/tags/form" %>
<html lang="en">
<head>
    <meta charset="UTF-8"/>
    <title>Form Example - Register a Person</title>
    <link rel="stylesheet"
type="text/css" href="/webjars/bootstrap/4.3.1/css/bootstrap.min.css"/>

    <spring:url value="/css/main.css" var="springCss"/>
    <link href="${springCss}" rel="stylesheet"/>
</head>
<body>
    ...
    <main role="main" class="container">
        <h2>输入人员信息</h2>
        <form:form method="POST" action="/addPerson" modelAttribute="person" class=
```

```html
"needs-validation">
            <form:errors path="*" cssClass="errorbox" element="div"/>
            <table class="table.table-striped">
                <tr>
                    <td><form:label path="name">姓名</form:label></td>
                    <td><form:input path="name" class="form-control"/></td>
                    <td><form:errors path="name" cssClass="error" element="div"/></td>
                </tr>
                <tr>
                    <td><form:label path="email">E-mail</form:label></td>
                    <td><form:input type="email" path="email" class="form-control"/></td>
                    <td><form:errors path="email" cssClass="error"/></td>
                </tr>
                <tr>
                    <td><form:label path="dateOfBirth">出生日期</form:label></td>
                    <td><form:input type="date" path="dateOfBirth" class="form-control"/></td>
                </tr>
                <tr>
                    <td><form:label path="password">密码</form:label></td>
                    <td><form:password path="password" class="form-control"/></td>
                    <td><form:errors path="password" cssClass="error"/></td>
                </tr>
                <tr>
                    <td><form:label path="sex">性别</form:label></td>
                    <td>
                        男:<form:radiobutton path="sex" value="M"/><br/>
                        女:<form:radiobutton path="sex" value="F"/>
                    </td>
                </tr>
                <tr>
                    <td><form:label path="job">工作类型</form:label></td>
                    <td>
                        <form:radiobuttons items="${jobItem}" path="job" element="label class='radio'"/>
                    </td>
                </tr>
                <tr>
                    <td><form:label path="regin">所在城市</form:label></td>
                    <td>
                        <form:select path="regin" items="${reginItems}" class="custom-select d-block w-100"/>
                    </td>
                </tr>
                <tr>
                    <td><form:label path="book">喜爱书籍</form:label></td>
                    <td>
                        <form:select path="book" class="custom-select d-block w-100">
                            <form:option value="-" label="--请选择--"/>
                            <form:options items="${books}"/>
```

```
            </form:select>
        </td>
    </tr>
    <tr>
        <td><form:label path="fruit">喜爱水果</form:label></td>
        <td>
            <form:select path="fruit" items="${fruit}" multiple="true" class="custom-select d-block w-100"/>
        </td>
    </tr>
    <tr>
        <td><form:label path="receiveNewsletter">接收邮件通知</form:label></td>
        <td><form:checkbox path="receiveNewsletter"/></td>
    </tr>
    <tr>
        <td><form:label path="hobbies">爱好</form:label></td>
        <td>游泳：<form:checkbox path="hobbies" value="游泳"/><br>
            打球：<form:checkbox path="hobbies" value="打球"/><br>
            逛街：<form:checkbox path="hobbies" value="逛街"/>
        </td>
    </tr>
    <tr>
        <td><form:label path="favouriteLanguage">喜爱编程语言</form:label></td>
        <td>
            <form:checkboxes items="${favouriteLanguageItem}" path="favouriteLanguage" element="label class='radio'"/>
        </td>
    </tr>
    <tr>
        <td><form:label path="notes">备注</form:label></td>
        <td><form:textarea path="notes" rows="3" cols="20" class="form-control"/></td>
    </tr>
    <tr>
        <td><form:hidden path="id" value="43564"/></td>
    </tr>
    <tr>
        <td colspan="2"><input type="submit" value="提交" class="btn btn-primary btn-lg btn-block"/></td>
    </tr>
</table>
</form:form>
<br>
<spring:escapeBody htmlEscape="true">
<h1>Hello zioer</h1>
</spring:escapeBody>

<spring:escapeBody javaScriptEscape="true">
<script type="text/javascript">alert('hello zioer')</script>
```

```
        </spring:escapeBody>
    </main>
    <script type = "text/javascript"
        src = "webjars/jquery/3.4.1/jquery.min.js"></script>
    <script type = "text/javascript"
        src = "webjars/bootstrap/4.3.1/js/bootstrap.min.js"></script>
</body>
</html>
```

以上示例代码中,重点在 form:form 标签中内容,用到前面所讲 Spring 标签库中的大多数内容,最后,演示 spring:escapeBody 标签如何使用,结合 bootstrap 样式的使用,使得页面更加美观,效果如图 3.13 所示。

图 3.13 Spring 标签库示例

如果输入错误,将有提示,如图 3.14 所示。

图 3.14 错误提示

如图 3.14 所示,错误提示有两种方式,在实际开发中,可根据需要进行选择使用。

最后一个页面是展示页,比较简单,应用 JSTL 中的标签,在此不再重复讲述。项目完成并启动后,在浏览器中输入下面地址进行访问:

http://127.0.0.1:8080/home

本章小结

本章内容较多，包括 Thymeleaf、FreeMarker 和 JSP，理论较强，但为了讲解不枯燥，在介绍时，同时加入了示例代码便于理解，同时在各个知识点介绍结束后，介绍了一个较完整示例，以将相关知识点进行串联，更加便于理解和掌握。关于模板，并不止本章所介绍，其实还包括很多其他技术。JSP 技术是目前争议较大的一个技术，基于其发展较早，技术更新很快，有很多新的前端页面技术涌现，很多人建议摒弃 JSP，拥抱更新技术。其实，这应该谨慎对待，JSP 还是具有很多其他前端技术不具备的优点。结合本章案例，将有助于知识点的掌握。

第4章 数据操作——使用Spring JDBC

本章介绍在 Spring Boot 2 中使用 JDBC、JDBCTemplate 连接数据库操作。同时,为了简化操作和配置的复杂性,本章操作将基于内存数据库 H2,同时介绍 H2 数据库基本知识。

4.1 H2 数据库

为了简化数据库配置,先简单介绍 H2 数据库,以便后续章节讲解。涉及其他数据库操作,比如 MariaDB、MongoDB 等,将在后面章节专门进行讲解。在此,为什么先行介绍热衷于 H2 数据库?

H2 数据库是 Java 编写的、开源的关系型数据库,同时也是嵌入式引擎数据库,不受平台限制,运行极其方便,具有下面优点:
- 开源、运行速度非常快,采用 JDBC API 连接;
- 具有嵌入式、服务器两种模式,同时是内存数据库;
- 基于浏览器的 Console 应用;
- 体积小:只有大约 2 MB 的 jar 文件大小。

在 Java 项目中,几乎不用额外安装任何数据库,便可在项目中自启动,非常适合项目测试、案例讲解等。当然,作为内存数据库家族,还有其他类似的内存数据库,比如 Derby、HSQLDB 等,也非常优秀。

H2 数据库可以作为内存数据库,或是单独的 Server 模式运行。单独运行时,采用下面方式。
进入 H2 网站的下载页面:

> http://www.h2database.com

在下载页面中,提供了 H2 不同版本的多种下载方式,在这里,一般下载最新版本的 Zip 包,下载后解压缩进入其中的 bin 目录,如图 4.1 所示。

该目录下是编译后的 jar 文件以及运行脚本,如果是 Windows 系统,运行 h2.bat 或 h2w.bat 文件,如果是 Linux 系统,运行 h2.sh 脚本。在 Windows 系统下,运行 h2.bat 文件后,或在命令提示符下,进入 bin 目录后,输入下面命令运行:

名称	修改日期	类型
h2-1.4.199.jar	2019/3/13 14:58	Executable Jar File
h2.sh	2019/3/13 14:58	Shell Script
h2w.bat	2019/3/13 14:58	Windows 批处理...
h2.bat	2019/3/13 14:58	Windows 批处理...

图 4.1　H2 数据库文件

```
Java -cp h2-1.4.199.jar org.h2.tools.Server -web -webAllowOthers -tcp -tcpPort 19200
-tcpAllowOthers
```

运行成功后，能看到运行 IP 和端口，如图 4.2 所示。

图 4.2　运行 H2 数据库

在浏览器中输入如图 4.2 所示地址，包括端口，打开连接界面，如图 4.3 所示。

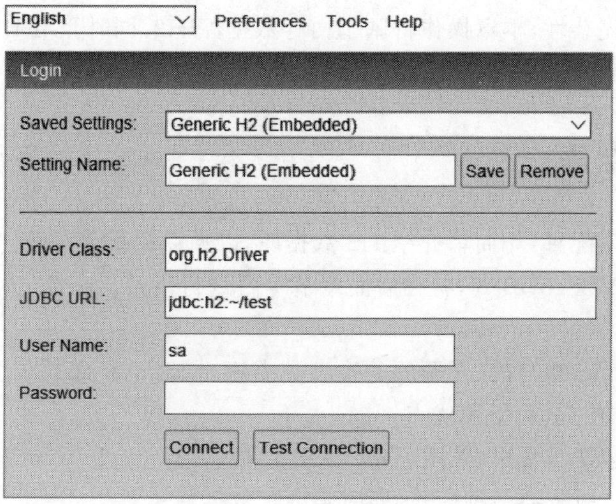

图 4.3　H2 登录界面

在该界面中，可更改界面显示语言为中文，编辑 JDBC URL 项，默认为当前用户目录下的 test 库，比如在 Windows 下数据库的位置为：

```
C:\Users\Li
```

Li 表示当前 Windows 登录用户，数据库名称为：test.mv.db。

Password 项不填写，表示密码默认为空，单击 connect 按钮可进入管理界面。在管理界面，可输入 SQL 进行操作，包括数据表的创建、删除、表数据的增、删、改和查询等操作。

提示：H2 数据库运行默认端口是 8082，运行前，保证该端口不被占用。或在上述命令之后加入参数-webPort，更改端口，如下所示：

```
Java -cp h2-1.4.199.jar org.h2.tools.Server -web -webAllowOthers -tcp -tcpPort 19200
-tcpAllowOthers -webPort 8083
```

以上命令将 Web 访问的端口号更改为 8083。

4.2 Java 连接 H2 数据库

Java 连接 H2 数据采用 JDBC 进行连接,分别有以下三种方式:
(1) TCP/IP 连接方式
H2 服务需要先启动,基于服务进行连接,多个客户端可同时连接到 H2 数据库。连接语法如下:

```
jdbc:h2:tcp://<server>[:<port>]/[<path>]<databaseName>
```

例如:

```
jdbc:h2:tcp://192.168.137.1:19200/~/test
```

注意:以上端口号一定要填写 TCP 端口号,而不是 Web 端口号。
(2) 嵌入式(本地)连接方式
以本地文件名方式进行连接,即独占文件的方式。连接语法如下所示:

```
jdbc:h2:[file:][<path>]<databaseName>
```

例如:

```
jdbc:h2:~/test                    //当前用户目录下的 test 数据库
jdbc:h2:file:/data/test           //Linux 下,使用路径
jdbc:h2:file:C:/Users/Li/test     //Windows 下,使用路径
```

注意:以此方式连接 H2 数据库,一次只能有一个连接打开,否则将报错。
(3) 内存模式
此方式支持直接在内存中创建数据库和表,但是,只在程序运行期间,所有信息都保存在内存。一旦程序关闭或重启,所有数据将不存在。连接语法如下所示:

```
jdbc:h2:mem:db
```

Java 采用 JDBC,以上三种方式都可以连接到 H2 数据库,其各有优缺点。前两种方式优点是所有数据保存在硬盘中,数据不会丢失,但是运行速度取决于硬盘速度快慢。第一种方式的缺点是需要单独启动 H2 服务,并占用端口,优点是可以多客户端连接,即 H2 数据库不一定放在本地计算机中;第二种方式优点是不需要单独启动服务,只需要知道 H2 数据库位置即可,缺点是必须以独享模式打开数据库,即一次只能有一个客户端连接;第三种方式的缺点是数据只存在于程序运行期间,优点是其运行于内存中,即速度快,适合于程序测试,可保证程序运行初始时,数据永远是干净的。

4.3 Spring Boot 2 中 JDBC 连接方式

JDBC(Java Data Base Connectivity),即 java 数据库连接,是用于执行 SQL 语句的 Java API,为多种关系型数据库提供统一访问方式,由一组用 Java 语言编写的类和接口组成。

JDBC 提供了标准的 API,可以构建更高级的工具和接口。

JDBC 具有如下优点:
- 接近访问数据库底层,访问速度快,执行效率高;
- 更安全,能处理大量数据信息;
- 开发人员能使用原生 SQL 处理数据。

同时,具有如下缺点:
- 重复代码量大;
- 操作相对烦琐,维护复杂;
- 开发人员掌握知识点多,包括写大量 SQL 代码。

基于 JDBC 所具有的缺点,特别是在中大型业务系统中,开发人员需要书写大量 SQL 语句,以及很多重复代码。于是诞生了很多面向对象以及封装后第三方框架,提高开发效率,比如著名的 Hibernate、MyBatis 等。这些封装后的框架至少在访问数据库速度上会有所下降,但在开发效率上提供了很大方便。但基于 JDBC 重要性,很多遗留系统采用 JDBC,以及 Spring Boot 2 对 JDBC 的支持,在此介绍在 Spring Boot 2 中 JDBC 的操作。

在 Spring Boot 2 项目中使用 JDBC 的方法很简单,在 pom.xml 中加入相关依赖:

```
spring-boot-starter-jdbc
```

以及使用数据库的驱动依赖,比如 H2 数据库依赖,示例代码如下所示:

```xml
<dependency>
    <groupId>org.springframework.boot</groupId>
    <artifactId>spring-boot-starter-jdbc</artifactId>
</dependency>
<dependency>
    <groupId>com.h2database</groupId>
    <artifactId>h2</artifactId>
</dependency>
```

通过以上依赖,自动解决系统中所需 jar 包,这种方式加入的连接池默认是 HiKariCP,该连接池是 Spring Boot 2 推荐并默认加入,目前,通过实验数据比较,HiKariCP 性能最优。在代码中,有两种方式使用 JDBC:一种方式是直接连接,另一种方式是采用 PreparedStatement 方式。

(1) 采用直接连接方式示例代码如下:

```java
public void getConnection(){
    try {
        Class.forName("org.h2.Driver");
        connection = DriverManager.getConnection("jdbc:h2:mem:db", "SA", "");
        connection.setAutoCommit(false);
    } catch (Exception e) {
        e.printStackTrace(System.out);
    }
}
```

以上直接连接方式优点是当使用时才连接数据库,不使用时可断开连接,释放资源。如果系统访问数据库量小时,此方式较佳。如果访问数据库量大,并发也较多时,此方式不一定适用,此时,最好采用连接池方式,可节省资源,示例代码如下所示:

```
@Autowired
DataSource dataSource;                    //声明 DataSource

public void getConnection2() throws SQLException{
    connection = dataSource.getConnection();
    connection.setAutoCommit(false);
}
```

接着,在配置文件 application.properties 中配置相关信息,示例代码如下所示:

```
spring.datasource.url = jdbc:h2:mem:h2test;DB_CLOSE_DELAY = -1;DB_CLOSE_ON_EXIT = FALSE
spring.datasource.platform = h2
spring.datasource.username = sa
spring.datasource.password = 
spring.datasource.driverClassName = org.h2.Driver
```

通过以上步骤,完成使用连接池方式的 JDBC 配置。下面,可以直接在代码中使用 JDBC 操作数据库,示例代码如下:

```
public void createTables(){
    try {
        String createSQL = "create table STUDENT (ID int(11) NOT NULL auto_increment, NAME VARCHAR(45), ADDRESS VARCHAR(55),PRIMARY KEY (id))";
        connection.createStatement().executeUpdate(createSQL);
        System.out.println("Tables Created!!!");
    } catch (SQLException e) {
        e.printStackTrace(System.out);
    }
}
```

以上代码通过连接语句创建数据表。下面是插入数据代码:

```
public void useStatement(){
    try {
        String insertStudentSQL = "INSERT INTO STUDENT( NAME, ADDRESS) VALUES ('%s','%s');";
        Statement statement = connection.createStatement();

        statement.execute(String.format(insertStudentSQL, "Lucy","Peking"));
        statement.execute(String.format(insertStudentSQL, "Tom","ShanDong"));

        connection.commit();
    } catch (Exception e) {
        try {
            connection.rollback();
        } catch (SQLException ex) {
            System.out.println("Error during rollback");
            System.out.println(ex.getMessage());
        }
        e.printStackTrace(System.out);
    }
}
```

以上代码通过 statement 方式在数据表中直接插入数据。

（2）采用 PreparedStatement 方式，示例代码如下所示：

```java
public void usePreparedStatement(){
    try {
        String insertStudentSQL = "INSERT INTO STUDENT( NAME, ADDRESS) VALUES (?,?);";
        PreparedStatement studentStmt =
connection.prepareStatement(insertStudentSQL);

        studentStmt.setString(1,"Mary");
        studentStmt.setString(2,"YunNan");

        studentStmt.executeUpdate();
        connection.commit();
    } catch (Exception e) {
        try {
            connection.rollback();
        } catch (SQLException ex) {
            System.out.println("Error during rollback");
            System.out.println(ex.getMessage());
        }
        e.printStackTrace(System.out);
    }
}
```

以上两种插入数据方式都可行，但第一种方式直接执行插入 SQL 语句，其缺点是不能保证插入数据是否具有破坏性，第二种方式采用占位方式，可以防止 SQL 注入，安全性更高。建议在和用户交互需要数据库数据插入、更新等操作时，采用第二种方式更加安全。

下面示例代码是查询语句：

```java
public void useSelect() {
    try {
        String selectSql = "SELECT * FROM STUDENT";
        Statement statement = connection.createStatement();
        ResultSet resultSet = statement.executeQuery(selectSql);
        while(resultSet.next()) {
            System.out.print(resultSet.getString("ID"));
            System.out.print("  " + resultSet.getString("NAME"));
            System.out.println("  " + resultSet.getString("ADDRESS"));
        }
    }catch(Exception e) {
        System.out.println(e.getMessage());
    }
}
```

查询方式采用的是 executeQuery，JDBC 中其他操作语句 Update、Delete 等和上面操作类似。为了对一批数据快速处理，Statement 提供了 executeBatch 方法，可以同时对一批数据进行处理，比如下面示例代码：

```java
public void useStatementBatch(){
```

```
        try {
            String insertStudentSQL = "INSERT INTO STUDENT( NAME, ADDRESS) VALUES ('%s','%s');";
            Statement statement = connection.createStatement();

            statement.addBatch(String.format(insertStudentSQL, "Masa","Xian"));
            statement.addBatch(String.format(insertStudentSQL, "Kate","TianJin"));
            statement.addBatch(String.format(insertStudentSQL, "Nanni","GuangZhou"));
            statement.addBatch(String.format(insertStudentSQL, "Harry","HongKong"));
            statement.addBatch(String.format(insertStudentSQL, "Yingk","NanNing"));

            statement.executeBatch();
            connection.commit();
        } catch (Exception e) {
            try {
                connection.rollback();
            } catch (SQLException ex) {
                System.out.println("Error during rollback");
                System.out.println(ex.getMessage());
            }
            e.printStackTrace(System.out);
        }
    }
```

以上代码是对插入操作的批量处理，特别是数据量大时，操作速度非常快。详细代码请查看本章示例代码。JDBC 相关代码掌握起来非常容易，但由于其书写代码量大，故出现很多基于 JDBC 的封装框架。下面介绍的是 Spring JDBCTemplate 框架。

4.4 Spring JDBCTemplate

SpringJDBCTemplate 是 Spring 对数据库的操作在 JDBC 上面做了深层次的封装，使用 Spring 的注入功能，在原始 JDBC 基础上，构建了抽象层，提供了许多使用 JDBC 的模板和驱动模块，消除了重复的 JDBC 代码，使操作数据库变得更简单，为 Spring 应用操作关系数据库提供了便利。

在 Spring Boot 2 中，使用 Spring 的注入功能，可以把 DataSource（数据库连接池）注册到 jdbcTemplate 中，示例代码如下：

```
@Autowired
private JdbcTemplate jdbcTemplate;
```

JDBCTemplate 主要提供了下面几类方法：

- query 方法及 queryForObject 方法：用于执行查询相关语句；
- update 方法及 batchUpdate 方法：update 方法用于执行单次增、删和改等语句；batchUpdate 方法用于执行批处理相关语句；
- execute 方法：可用于执行任何 SQL 语句，多用于执行操作数据库 DDL 语句；
- call 方法：可用于执行存储过程、函数相关语句。

下面是相关示例详细介绍。

建立一个新的 Spring Boot 2 工程,编辑 pom.xml 文件,增加如下依赖:

```xml
<dependency>
    <groupId>org.springframework.boot</groupId>
    <artifactId>spring-boot-starter-jdbc</artifactId>
</dependency>
<dependency>
    <groupId>com.h2database</groupId>
    <artifactId>h2</artifactId>
</dependency>
```

由上面代码可知,使用 Spring JDBCTemplate 时,只需要加入 spring-boot-starter-jdbc 依赖即可,便可加入所有相关依赖,包括默认连接池等,h2 提供 H2 驱动依赖。在一个简单工程中,以上两个依赖完成了关于 JDBCTemplate 和 H2 数据库的 Jar 包管理。

接着,编写配置文件 application.properties,加入下面内容:

```
spring.datasource.url = jdbc:h2:mem:h2test;DB_CLOSE_DELAY = -1;DB_CLOSE_ON_EXIT = FALSE
spring.datasource.platform = h2
spring.datasource.username = sa
spring.datasource.password =
spring.datasource.driverClassName = org.h2.Driver
spring.datasource.schema = classpath:db/schema.sql
spring.datasource.data = classpath:db/data.sql

spring.h2.console.enabled = true
spring.h2.console.path = /console
spring.h2.console.settings.trace = false
spring.h2.console.settings.web-allow-others = false
```

以上包含两部分,一个是 spring dataSource 相关配置,这是主要配置,其中包含了 schema 和 data 的设置,表示系统在启动时,运行指定路径下的数据库初始脚本,包括表的创建和初始数据。但实际上,如果不进行 schema 和 data 的设置,将 schema.sql 和 data.sql 放在 /src/main/resources 路径下,系统启动时,同样可以自动运行。在这里,将数据表创建语句放在 schema.sql 中,内容如下:

```sql
CREATE TABLE student(
 ID int(11) NOT NULL auto_increment,
 NAME varchar(20) null,
 AGE int(11) null,
 PRIMARY KEY (id)
);
```

文件中只放了一个创建数据表 student 的语句。将初始数据放在文件 data.sql 中,示例如下所示:

```sql
INSERT  INTO  student ( NAME, AGE) VALUES ('张三',10);
INSERT  INTO  student ( NAME, AGE) VALUES ('李四',16);
```

介绍以上方法的优点是,如果系统处于开发和测试阶段,可以将所有创建表和初始数据放在上面两个文件中，系统一旦启动，便重新初始化，方便系统测试。配置方法只需要用以上介绍内容便轻松完成。创建 model 类文件 Student.java，用来描述学生信息，在后面 jdbcTemplate 调用中将使用到：

```java
public class Student {
    private int id;
    private String name;
    private int age;

    //省略 get 和 set

    @Override
    public String toString() {
        return "Student{" +
                "id = " + id +
                ", name = '" + name + '\'' +
                ", age = '" + age + '\'' +
                '}';
    }

}
```

以上创建模型代码比较简单，只是为了示例说明，在代码最后，覆盖重写了 toString() 方法，为了在打印该类时，具有更直观可调试性。在实际开发中，可改写该方法，返回为 JSON 格式等。创建资源类文件 StudentDao.java，将所有和数据表 student 交互代码放置在文件中，便于管理：

```java
@Repository
public class StudentDao {
    …
}
```

创建类 StudentDao，并将其注解为@Repository，让系统启动扫描时，自动注册该类为资源类。接着，加入自动装配 JdbcTemplate，减少代码量，并增加可读性，同时不需要过多的人为干预，即可完成 JdbcTemplate 的实例化：

```java
@Autowired
private JdbcTemplate jdbcTemplate;
```

下面在类 StudentDao 创建各个方法，调用 jdbcTemplate：

```java
public void create(){
    String sql = "create table teacher (id int(11) NOT NULL auto_increment,name varchar2(40),age int(11)) ";
    jdbcTemplate.execute(sql);
    log.info("Table created!");
}
```

上面方法通过调用 execute，实现数据表的创建。下面方式是查询数据记录：

```java
public void useSelect() {
```

```java
    String sql = "SELECT * FROM student ";

    List<Map<String, Object>> users = jdbcTemplate.queryForList(sql);
    System.out.println(users);

    List<Student> queryAllList = jdbcTemplate.query(sql, new Object[]{},
            new BeanPropertyRowMapper<Student>(Student.class));
    log.info("查询学生情况:" + queryAllList);
}
```

在上面代码中，实现了两种查询记录方式，一是调用 queryForList，返回 List<Map<String，Object>>，另一种方式是调用 query，在其中使用 BeanPropertyRowMapper，实现数据库表字段和实体类自动对应关系，减少手动绑定工作量，有效提高开发效率，返回 List<Student>。以上两种方式均可达到返回 list 的作用。

```java
public void useSelectOne() {
    String sql = "SELECT * FROM student where ID = ?";

    RowMapper<Student> rowMapper =
        new BeanPropertyRowMapper<>(Student.class);
    Student student = jdbcTemplate.queryForObject(sql, rowMapper, 1);

    log.info("查询一个学生情况:" + student);
}
```

以上代码中，调用 queryForObject，实现单条记录的返回。注意其中参数，第一个参数是需要查询的 SQL 语句，第二个参数是使用 BeanPropertyRowMapper，实现数据库表字段和实体类自动对应关系，第三个及以后参数是 SQL 中占位符的参数值，如果没有，则可以不存在，如果有多个，则会存在更多参数。最后返回的是实例 student。

```java
public void useSelectcount() {
    String sql = "SELECT count(1) FROM student where id>?";
    long count = jdbcTemplate.queryForObject(sql, Long.class, 1);

    log.info("查询行数:" + count);
}
```

上面代码实现如何调用 queryForObject 只返回一个值的情况，注意其中第二个参数为 Long.class。

```java
public void useSelectOneByRowCallback(){
    Student student   = new Student();

    //该方法返回值为 void
    this.jdbcTemplate.query("select * from student where id = ?",
            new Object[] { 10 },
            new RowCallbackHandler() {

                @Override
                public void processRow(ResultSet rs) throws SQLException {
```

```java
                student.setId(rs.getInt("id"));
                student.setName(rs.getString("name"));
                student.setAge(rs.getInt("age"));
            }
        });

        log.info("RowCallbackHandler 查询: " + student);
    }
```

上面代码通过 RowCallbackHandler，以及手工对应方式，返回一条记录，其中类 processRow 只使用一次，这里采用内部类的方式实现，减少实例化操作。同理，也可返回多条记录。

```java
    public void useUpdate() {
        String sql = "UPDATE student SET NAME = ? WHERE ID = ?";
        jdbcTemplate.update(sql, "Jack", 2);
        log.info("更新学生情况 ");
    }

    public void useBatchUpdate() {
        String sql = "INSERT INTO student ( NAME, AGE) VALUES(?,?)";

        List<Object[]> batchArgs = new ArrayList<>();

        batchArgs.add(new Object[]{"Jula", 11});
        batchArgs.add(new Object[]{"Kema", 22});
        batchArgs.add(new Object[]{"Tiny", 13});
        batchArgs.add(new Object[]{"Yisuy", 13});
        batchArgs.add(new Object[]{"Pybing", 22});

        jdbcTemplate.batchUpdate(sql, batchArgs);
        log.info("批量增加学生 ");
    }

    //避免 SQL 注入
    public void useUpdate2() {
        jdbcTemplate.update("insert into student(name,age) values(?,?)",
            new PreparedStatementSetter(){
                @Override
                public void setValues(PreparedStatement ps) throws SQLException {
                    ps.setString(1, "huloo");
                    ps.setInt(2, 31);
                }
            });
        log.info("避免 SQL 注入方式增加学生");
    }
```

上面三个方法，分别实现单条记录更新、批量插入和避免 SQL 方式的插入。调用了 update、batchUpdate 和 PreparedStatementSetter。注意其中的调用方法。

```java
    public void useDelete() {
```

```
        String sql = "DELETE student WHERE ID = ?";
        jdbcTemplate.update(sql, 4);
        log.info("删除一个学生 ");
}
```

上面代码中,调用 update 实现了删除一条记录。

通过上面代码介绍,使用 Spring JDBCTemplate,实现数据库操作时,减少在调用 SQL 语句时冗余代码的书写,提高代码的可读性。但是,也要看到其中弊端,SQL 语句暴露在 Java 语句中,尽管开发人员可以将其放置在一个或多个相关类中集中管理,但在后期 SQL 的维护中,将带来一定工作量,特别是在大型系统中。以上示例工程目录结构如图 4.4 所示,详细代码请查看本章源码。

图 4.4　示例工程目录

本章小结

本章介绍了数据操作的两种方式,一种是在 Spring Boot 2 中如何调用原生 JDBC,另一种是介绍如何调用 Spring JDBCTemplate。这两种方式都很接近底层访问数据方式,其在实际开发中占有一定重要地位,特别注意的是,在 Spring Boot 2 中,配置已变得很简单,开发人员几乎无须关心如何管理依赖。接近底层访问数据方式可使得访问数据库速度非常快,因这个特点,使得其开发效率相对低下、后期维护变得复杂。故很多开发人员致力于封装 JDBC,有的侧重于开发方式改变,比如 Spring JDBCTemplate,很接近原生 JDBC,但开发效率得到的提高不是很大,另一些则致力于改变开发方式,比如面向对象方式,还有些致力于多方面兼顾,等等。

第5章
数据操作——Spring Data JPA

Spring JPA 是 Spring Data JPA 的简称，可理解为 Spring 整合 JPA 的东西。Spring 是包容性很强大的容器，整合 JPA 实际上是简化关系数据库操作。本章将介绍 JPA，以及 Spring 中操作 JPA 的方式和案例。

5.1 JPA 介绍

JPA 是什么？JPA，即 Java Persistence API 缩写，意思是 Java 持久层 API。这就好理解了，JPA 是和数据库打交道那一层，将运行期的实体对象持久化到数据库中。目的是简化应用系统开发中的开发工作量。

JPA 是 JCP 组织（Java Community Process，一个开放的国际组织）发布的 Java EE 标准之一，只要符合 JPA 标准的框架都使用相同的架构，提供统一访问 API，保证基于 JPA 开发的应用系统能只需少量修改就可以适应不同的 JPA 框架并运行。

JPA 支持大数据集、事务、并发等事务。其目标之一是提供简单编程模型：保证在 JPA 中创建实体和创建 Java 类同样方便与简单，即没有约束和限制，只需使用 javax.persistence.Entity 进行注释；其次，其接口也很简单，开发者能很容易地掌握。JPA 思想是基于非侵入式原则设计，以便容易地和其他框架或者容器集成。

JPA 查询语言是面向对象，提供面向对象的自然语法构造查询语句，其定义的 JPQL（Java Persistence Query Language）是 EJB QL 的一种扩展，其是针对实体的一种查询语言，操作对象是实体，而非关系数据库的表，即通过类名和属性访问，而不是表名和表的属性。同时，JPA 能支持面向对象的高级特性，包括类之间的继承、多态和类之间的复杂关系，可用于复杂的业务模型设计。

5.2 Spring Data JPA

从字面上可理解，Spring Data JPA 即是 JPA 的实现、封装和再次抽象。底层使用了 Hibernate 的 JPA 技术实现。Hibernate 是一个开放源代码的对象关系映射框架，其对

JDBC进行了对象封装,它将POJO与数据库表建立映射关系,可理解为全自动的ORM框架。

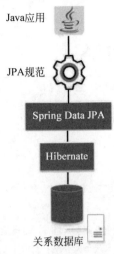

图5.1 Spring Data JPA 和 Hibernate 关系

前面章节介绍了JDBC,其最大不便之处是将暴露大量SQL语句在程序代码中,那么ORM封装了SQL语句,只将业务模型面向开发人员,开发人员不需关心如何和数据库打交道,以及SQL实现。Hibernate将对数据库的操作转换为对Java对象的操作,屏蔽不同数据库实现之间的差异,因此简化开发。简单理解Spring Data JPA和Hibernate等其他框架关系如图5.1所示。

通过图5.1,可理解Spring Data JPA是在Hibernate基础上再封装一层,具有很强扩展性,如果可能,将Hibernate替换为其他实现,Spring Data JPA也是可用的。由此,加深对Spring Data JPA的理解。

快速理解Spring Data JPA的最好方法是使用示例。本节介绍一个简单的基于Spring Boot 2的Spring Data JPA示例。新建Spring Boot 2项目,编辑pom.xml文件,增加如下依赖:

```xml
<dependency>
    <groupId>org.springframework.boot</groupId>
    <artifactId>spring-boot-starter-data-jpa</artifactId>
</dependency>
<dependency>
    <groupId>org.springframework.boot</groupId>
    <artifactId>spring-boot-starter-validation</artifactId>
</dependency>
<dependency>
    <groupId>com.h2database</groupId>
    <artifactId>h2</artifactId>
</dependency>
```

上面代码加入spring-boot-starter-data-jpa依赖,自动加入Spring Data JPA相关依赖,spring-boot-starter-validation依赖将加入验证相关依赖,h2依赖加入H2数据库驱动。

下面创建业务模型User.java,主要代码如下所示:

```java
@Entity//标识该类为一个实体
@Table(name = "user")//关联数据库中的user表
public class User {
    @Id
    @GeneratedValue(strategy = GenerationType.IDENTITY)
    private Long id;
    @NotNull
    private String name;
    private String address;
    private String phone;

    public User() {
```

```
    }

    public User(String name,String address,String phone) {
        this.name = name;
        this.address = address;
        this.phone = phone;
    }

    //省略 get 和 set...
}
```

上面代码中几个重要注解：
- @Entity：表明该类（User）为一个实体类，默认类名是数据库表中表名，该类中字段名即表中的字段名。默认系统初始化时，会在数据库中创建该表；
- @Table：该注解用来改变类名与数据库中表名的映射规则，如上代码所示，User 类映射数据库中表名 user；
- @Id：该注释用来将 Java 字段标记为数据库表主键列；
- @GeneratedValue：该注释用来表示生成主键列的策略；
- @NotNull：该注释用来表示 Java 字段对应的数据库表中列不能为 Null。

以上几个注解很重要，标识了如何创建数据表。下面创建 Dao 接口类，继承自 JpaRepository，代码如下所示：

```
public interface UserDao extends JpaRepository<User,Long> {
}
```

值得注意的是，代码就这么简单，这样就可以完成对数据表简单增加、删除、修改和查询等操作。接口 JpaRepository<T,ID>中有两个参数：
- T：表示需要类型化为实体类，比如 User 类；
- ID：表示实体类 T 中关键字 Id 的类型，比如上面 User 类中关键字 Id 的类型为 Long。

下面是操作数据表方法，创建服务层 UserService.java，用于操作数据表，示例代码如下所示：

```
@Service
public class UserService {
    @Autowired
    UserDao userDao;

}
```

上面代码创建时，标识 UserService 类为@Service，以让系统启动时，将该类注解为服务，在该类中定义全局变量 UserDao，并采用注解@Autowired，即可自动装配 UserDao。下面是该类中的一些方法：

```
public void useSave() {
    userDao.save(new User("Tom","Peking","13811111"));
    userDao.save(new User("Lucy","ShanXi","13612345"));
    userDao.save(new User("Funny","GuangZhou","13354321"));
```

```java
    }

    public void useFindAll() {
        for (User user : userDao.findAll()) {
            System.out.println(user.toString());
        }
    }

    public void useFindId() {

        userDao.findById(2L)
        .map(user ->{
            System.out.println("User found with findById(2):");
            System.out.println(user);
            return true;
        }).orElse( false );

    }

    public void useCount() {

        System.out.println("found count : " + userDao.count());
    }
```

在上面代码中,在 useSave()方法里调用了 save()方法,在上面定义中发现并没有定义该方法,其实,该方法继承自 Repository,其提供了最基本的数据访问功能,由此,可知 JPA 确实简化不少代码量,没有写一句 SQL 语句,便可完成数据表中数据的增加。

同理,在 useFindAll()方法里调用了 findAll()方法,用于返回整个数据表中记录;useFindId()方法中,使用了 findById()方法,用于查找指定关键字记录;useCount()方法中,使用了 count()方法,用于统计数据表中记录数。

接着,在适当地方调用上面方法,进行测试,主要代码如下所示:

```java
@Autowired
UserService userService;

userService.useSave();                  //保存
userService.useFindAll();               //查找全部
userService.useFindId();                //查找指定 Id 值
userService.useCount();
```

至此,完成示例代码编写,其完整数据结构如图 5.2 所示。

以上代码特点是,没有作任何配置和显式创建数据表,只创建了模型,JPA 根据该模型自动创建数据表 user;没有编写 SQL 语句,dao 层继承了接口 JpaRepository,便轻松实现数据表的操作。由此可知,JPA 确实简化了数据表的操作,并能很大程度提高代码开发效率。

以上完整示例请详见本章源码。

```
v 🗁 c5-1 [boot]
  v 🗁 src/main/java
    v ⊞ com.zioer
      v ⊞ dao
        > 🗋 UserDao.java
      v ⊞ model
        > 🗋 User.java
      v ⊞ service
        > 🗋 UserService.java
      v 🗋 C51Application.java
        > ⓒ C51Application
  v 🗁 src/main/resources
    📄 application.properties
  > 🗁 src/test/java
  > 🗁 JRE System Library [JavaSE-1.8]
  > 🗁 Maven Dependencies
  v 🗁 src
    > 🗁 main
    > 🗁 test
  > 🗁 target
    📄 pom.xml
```

图 5.2　示例代码结构

5.3　JpaRepository＜T，ID＞方法

　　5.2 节介绍了 dao 层只需继承自接口 JpaRepository，便可快速实现数据库表的基本操作，本节将介绍接口 JpaRepository 提供的所有方法，以便在开发中灵活运用。

　　查看 JpaRepository 源码，其提供了如下方法：

　　(1) List＜T＞ findAll()；//查询所有记录

　　该方法用于查询数据表中所有记录，一般数据量比较大或有条件查询时，不建议使用该方法。

　　(2) List＜T＞ findAll(Sort sort)；//用排序方式查询所有记录

　　和上一个方法区别是，需要预先指定排序方式，然后查询所有记录。排序定义方式如下：

```
List<Sort.Order> list = new ArrayList<>();
Order order1 = new Order(Direction.ASC, "name");

list.add(order1);

Sort sort = Sort.by(list);
```

　　在上面定义 sort 方法中，可以预先指定多字段排序方式，然后在 findAll() 方法中作为参数，示例如下所示：

```
userDao.findAll(sort);
```

　　(3) List＜T＞ findAllById(Iterable＜ID＞ ids)；//根据关键字 id 集合，返回记录 list 示例如下所示：

```java
Collection<Long> a = new ArrayList<Long>();            //定义一个Long集合
a.add(2L);
a.add(4L);
//调用findAllById,并打印输出
for (User user : userDao.findAllById(a)) {
    System.out.println(user.toString());
}
```

(4) <S extends T> List<S> saveAll(Iterable<S> entities);//批量保存

示例如下所示：

```java
Collection<User> users = new ArrayList<User>();        //定义User的集合
//加入多个User实例
users.add(new User("MingMing","ZhenZhou","13811651"));
users.add(new User("TianTian","ShangHai","13612875"));
users.add(new User("DanDan","AoMen","13358611"));
//调用saveAll方法
userDao.saveAll(users);
```

(5) void flush();//立即写入数据库

(6) <S extends T> S saveAndFlush(S entity);//保存数据并且立即写入数据库

(7) void deleteInBatch(Iterable<T> entities);// 批量删除

示例如下：

```java
Collection<User> userList = new ArrayList<User>();  //创建User集合
//加入User,注意每个User只需要关键字id值即可
userList.add(new User(6L));
userList.add(new User(8L));
userDao.deleteInBatch(userList);                    //调用deleteInBatch方法进行删除
```

(8) void deleteAllInBatch();//删除批量调用中的所有实体,实际是清空表

示例如下：

```java
userDao.deleteAllInBatch();                         //清空表user
```

(9) T getOne(ID id);//根据id获得一个对象引用

在实际使用中,不建议使用该方法。

(10) <S extends T> List<S> findAll(Example<S> example);//根据example查询

示例如下：

```java
User u = new User();
u.setAddress("ShanXi");
Example<User> example = Example.of(u);              //构造Example实例
for (User user : userDao.findAll(example)) {
    System.out.println(user.toString());
}
```

(11) <S extends T> List<S> findAll(Example<S> example, Sort sort);//根据条件查询,两个参数,一个是example,另一个是排序规则sort

示例如下：

```
User u = new User();
u.setAddress("ShanXi");
Example<User> example = Example.of(u);         //构造 example
List<Sort.Order> list = new ArrayList<>();
Order order1 = new Order(Direction.DESC, "name");

list.add(order1);
Sort sort = Sort.by(list);                     //构造 sort

for (User user : userDao.findAll(example,sort)) {
    System.out.println(user.toString());
}
```

除了以上介绍方法外，JpaRepository 还继承自 PagingAndSortingRepository 和 QueryByExampleExecutor，还可以调用这两个类中其他方法，如下所示：

- `findOne(Example<S> example)`   //返回一个对象，注意和前面 getOne 区别
- `findAll(Example<S> example)`   //返回满足条件的所有对象
- `findAll(Example<S> example, Sort sort)`   //查询满足条件的所有对象，并带排序
- `findAll(Example<S> example, Pageable pageable)`   //分页查询满足条件的所有对象
- `long count(Example<S> example)`   //根据条件进行统计
- `boolean exists(Example<S> example)`   //根据条件，判断是否存在

以上仅列举部分方法和简单案例，但由此可窥，JPA 提供了众多方法供开发人员根据实际情况进行调用，以达到能集中于业务需求而快速开发的目的。以上案例可参考源码部分。

5.4　接口规范名方法

前面介绍 JPA 提供的方法基本能满足大部分需求，但需求是反复变化的，不能满足所有需求，或者需要写很多重复代码，由此，开发人员还是希望能简化代码。所以，Sping Data JPA 提供接口规范方法名查询方式，以进一步满足用户需求。

在创建的 Dao 层，默认接口定义如下所示：

```
public interface UserDao extends JpaRepository<User,Long> {
}
```

在该接口中加入下面方法名：

```
List<User> findByName(String name);
```

接着，在 Service 层调用该方法，如下代码所示：

```
public void usefindByName() {
    for (User user : userDao.findByName("Mary")) {
        System.out.println(user.toString());
    }
}
```

这样，可以查询出条件 name="Mary"对应的所有记录。

是不是很简单？没有写任何SQL语句，只是简单调用方法：

List<User> findByName(String name);

便完成了所需功能。这就是本节所介绍接口规范方法名查询。即只需在继承Repository接口的接口文件中使用JPA规范的方法命名，Spring Data JPA就可按照约定进行查询操作。

基本定义方法是：查询方法以find、read或get开头，涉及条件查询时，条件的属性用条件关键字连接，整个查询方法名建议按驼峰式命名，JpaRepository会对方法名进行校验，不符合规范会报错。在定义方法名时，需满足如表5.1所示规则。

表5.1 方法名称中支持的关键字

序号	关键字	示例	生成JPQL代码段
1	And	findByNameAndUrl	where x.name=?1 and x.url=?2
2	Or	findByNameOrUrl	where x.name=?1 or x.url=?2
3	Is,Equals	findByName findByNameIs findByNameEquals	where x.name=?1
4	Between	findByCreatetimeBetween	where x.createtime between ?1 and ?2
5	LessThan	findByNumLessThan	where x.num < ?1
6	LessThanEqual	findByNumLessThanEqual	where x.num <= ?1
7	GreaterThan	findByNumGreaterThan	where x.num > ?1
8	GreaterThanEqual	findByNumGreaterThanEqual	where x.num >= ?1
9	After	findByCreatetimeAfter	where x.createtime > ?1
10	Before	findByCreatetimeBefore	where x.createtime < ?1
11	IsNull	findByNumIsNull	where x.num is null
12	IsNotNull,NotNull	findByNum(Is)NotNull	where x.num not null
13	Like	findByNameLike	where x.name like ?1
14	NotLike	findByNameNotLike	where x.name not like ?1
15	StartingWith	findByNameStartingWith	where x.name like ?1（参数中需加入符号：%）
16	EndingWith	findByNameEndingWith	where x.name like ?1（参数中需加入符号：%）
17	Containing	findByNameContaining	where x.name like ?1（参数中需加入符号：%）
18	OrderBy	findByNumOrderByNameDesc	where x.num=?1 order by x.name desc
19	Not	findByNameNot	where x.name <> ?1
20	In	findByNumIn(Collection<Num> nums)	where x.num in ?1
21	NotIn	findByNumNotIn(Collection<Num> nums)	where x.num not in ?1
22	TRUE	findByStatusTrue()	where x.status=true
23	FALSE	findByStatusFalse()	where x.status=false
24	IgnoreCase	findByNameIgnoreCase	where UPPER(x.name)=UPPER(?1)

表 5.1 很清楚表明如何自定义构造一个方法，以及 JpaRepository 支持什么样的自定义构造方法，几乎支持一个查询语句中 where 相关关键字，包括 AND、OR、BETWEEN…AND、<=、>=、<、>、LIKE、NOT LIKE、ORDER BY 等。

另外，可在方法中加入特殊的参数，即直接在方法的参数上加入分页或排序的参数，例如下面代码：

```
Page findByUrl(String url, Pageable pageable);
List findByUrl(String url, Sort sort);
```

下面是相关示例，代码中有具体说明：

```
public interface UserDao extends JpaRepository<User,Long> {
    /**
     * 单条件查询
     * 对应 sql:select u from User u where u.name = ?1
     * 参数名大写,条件名首字母大写,并且接口名中参数出现的顺序必须和参数列表中的参数顺序一致
     */
    List<User> findByNameIs(String name);

    /**
     * and 条件查询
     * 对应 sql:select u from User u where u.name = ?1 and u.address = ?2
     * 参数名大写,条件名首字母大写,并且接口名中参数出现的顺序必须和参数列表中的参数顺序一致
     */
    User findByNameAndAddress(String name, String address);

    /**
     * or 条件查询
     * 对应 sql:select u from User u where u.name = ?1 or u.address = ?2
     */
    List<User> findByNameOrAddress(String name, String address);

    /**
     * between 查询
     * 对应 sql:select u from User u where u.birthday between ?1 and ?2
     */
    List<User> findByBirthdayBetween(Date startdate, Date enddate);

    /**
     * less 查询
     * 对应 sql:select u from User u where u.birthday < ?1
     */
    List<User> findByBirthdayLessThan(Date time);

    /**
     * greater 查询
     * 对应 sql:select u from User u where u.birthday > ?1
     */
```

```java
List<User> findByBirthdayGreaterThan(Date time);

/**
 * is null 查询
 * 对应 sql:select u from User u where u.name is null
 */
List<User> findByNameIsNull();

/**
 * Is Not Null 查询
 * 对应 sql:select u from User u where u.phone is not null
 */
List<User> findByPhoneIsNotNull();

/**
 * like 模糊查询
 * 这里的模糊查询并不会自动在 name 两边加"%",需要手动对参数加"%"
 * 对应 sql:select u from User u where u.name like ?1
 */
List<User> findByNameLike(String name);

/**
 * Not Like 模糊查询
 * 对应 sql:select u from User u where u.name not like ?1
 */
List<User> findByNameNotLike(String name);

/**
 * 倒序排序查询
 * 对应 sql:select u from User u where u.address = ?1 order by u.birthday desc
 */
List<User> findByAddressOrderByBirthdayDesc(String address);

/**
 * <>查询
 * 对应 sql:select u from User u where u.phone <> ?1
 */
List<User> findByPhoneNot(String phone);

/**
 * in 查询,方法的参数可以是 Collection 类型,也可以是数组或者不定长参数
 * 对应 sql:select u from User u where u.id in ?1
 */
List<User> findByIdIn(List<Long> ids);

/**
 * not in 查询,方法的参数可以是 Collection 类型,也可以是数组或者不定长参数
 * 对应 sql:select u from User u where u.id not in ?1
 */
List<User> findByIdNotIn(List<Long> ids);
```

```java
/**
 * 分页查询,方法的参数可以是 Collection 类型,也可以是数组或者不定长参数
 * 对应 sql:select u from User u where u.id not in ?1 limit ?
 */
Page<User> findByIdNotIn(List<Long> ids, Pageable pageable);

}
```

通过以上示例中的说明,可加深理解方法名称定义规则。在上面最后一个方法 findByIdNotIn()中,加入了分页参数"Pageable pageable"。即在调用时,用到数据分页显示。下面是调用示例:

```java
@Service
public class UserService {
    @Autowired
    UserDao userDao; //自动装配
    /**
     * 调用 findByNameAndAddress 方法
     */
    public void usefindByNameAndAddress() {
        System.out.println(" --- use findByNameAndAddress Example --- ");

        User user = userDao.findByNameAndAddress("Mary","NeiMengGu");
        System.out.println(user);
    }
    /**
     * 调用 findByBirthdayBetween 方法,注意日期传入方式
     */
    public void usefindByBirthdayBetween() throws ParseException {
        System.out.println(" --- use findByBirthdayBetween Example --- ");
        for (User user : userDao.findByBirthdayBetween(strTodate("2000-1-1"),strTodate("2000-6-1"))) {
            System.out.println(user);
        }
    }
    /**
     * 调用 findByNameLike 方法,当是 like 时,需传入符号%,根据需要,前置或后置
     */
    public void usefindByNameLike() {
        System.out.println(" --- use findByNameLike Example --- ");

        for (User user : userDao.findByNameLike("M%")) {
            System.out.println(user);
        }
    }
    /**
     * 调用 findByAddressOrderByBirthdayDesc 方法
     */
    public void usefindByAddressOrderByBirthdayDesc() {
        System.out.println(" --- use findByAddressOrderByBirthdayDesc Example --- ");
```

```java
        for (User user : userDao.findByAddressOrderByBirthdayDesc("ShangHai")) {
            System.out.println(user);
        }
    }
    /**
     * 调用 findByIdIn 方法,提前构造 List
     */
    public void usefindByIdIn() {
        System.out.println(" --- use findByIdIn Example --- ");
        List<Long> ids = new ArrayList<Long>();

        ids.add(3L);
        ids.add(4L);
        ids.add(7L);

        for (User user : userDao.findByIdIn(ids)) {
            System.out.println(user);
        }
    }
    /**
     * 调用 findByIdNotIn 方法,提前构造分页 Pageable
     */
    public void usefindByIdNotInpageable() {
        System.out.println(" --- use findByIdNotIn pageable Example --- ");
        List<Long> ids = new ArrayList<Long>();

        ids.add(3L);
        ids.add(4L);
        ids.add(7L);

        Pageable pageable = null;

        List<Sort.Order> list = new ArrayList<>();
        Order order1 = new Order(Direction.DESC, "id");

        list.add(order1);

        Sort sort = Sort.by(list);
        pageable = PageRequest.of(0, 2, sort);

        for (User user : userDao.findByIdNotIn(ids, pageable)) {
            System.out.println(user);
        }
    }
    /**
     * 自定义函数 strTodate,简化字符串转为日期
     */
    public Date strTodate(String str) throws ParseException {
        DateFormat format1 = new SimpleDateFormat("yyyy-MM-dd");
        return format1.parse(str);
    }

}
```

限于篇幅,以上代码只是部分示例。提示:将该类注解为服务后,便可调用前面自定义的方法,但一些特殊参数需要先构造,比如分页等。详细代码请看本节源码。

5.5 @Query

Spring Data JPA 提供了一种更自由、灵活的方式来构造 SQL 语句,与上面介绍的几种方法相比,它无需强制方法名,而是使用自己的一套规则,更加强大的是还能对数据进行更新。

此时,需要在继承 JpaRepository 的接口中使用注解@Query,完成用户定制化需求。

使用@Query 的好处包括可以控制参数方式、接口自定义、更加复杂的查询、同时支持使用 JPQL 语句和本地 SQL 语句,以及更新数据库表记录等操作。下面是示例代码:

```java
public interface UserDao extends JpaRepository<User,Long> {
    /**
     * 根据 address 查询
     */
    @Query("select x from User x where x.address = ?1")
    List<User> findUsersByAddress(String address);
}
```

以上代码中,自定义方法 findUsersByAddress()放在继承了 JpaRepository<>的接口 UserDao 中。在自定义方法上使用了注解@Query,即实现自定义查询数据表 user 记录。在以上 JPQL 语句使用了基于位置的参数绑定,比如"?1"替代参数,在开发中很容易出错。解决此问题的方法是,在语句中使用命名化参数,即使用@Param 表达,并在查询中绑定名称,示例代码如下所示:

```java
/**
 * 支持命名参数:根据 name 查询
 */
@Query("select x from User x where x.name = :name")
List<User> findUsersByName(@Param("name") String name);
```

在以上代码中,使用了注解@Param 来表达一个参数,即可以在@Query 语句中用类似于":name"来替代"?1",使得语句更具有可读性,需要注意其放置位置。

如果开发人员想使用 SQL 语句来直接操作数据库表,那么可以在@Query 语句中使用 nativeQuery 参数,并设置其为 true 即可,示例代码如下所示:

```java
/**
 * 本地查询
 */
@Query(value = "select * from user where name like CONCAT('%',?1,'%')", nativeQuery = true)
List<User> findUsersByNameLocal(String name);
```

以上代码示例如何使用 SQL 语句进行查询。

@Query 除了支持查询外，还支持对数据库表的更新，如果是更新操作，需要加上注解 @Modifying 和 @Transactional，否则运行时会报错，示例代码如下所示：

```
/**
 * 更新 phone
 */
@Transactional
@Modifying
@Query("update User x set x.phone = ?1 where x.id = ?2")
Integer updatePhoneById(String phone, Long id);
```

提示：在更新操作时，返回值只能是 int/Integer 或 void。

本节介绍了功能强大的 @Query 注解，使用该注解可灵活构建查询语句，以及更新语句操作。

@Query 注解在当前版本的本地查询不支持动态排序和翻页。

5.6 多表查询

多表查询指表与表之间关系。在具体业务的数据库设计中，多表关系很常见，其能表达更多内容，比如学生信息和选课信息就涉及至少两张表之间关系，需要考虑两张表之间映射关系，更多张表时，可以类推。两表之间最常见的映射关系包括一对多映射、一对一映射和多对多映射关系。下面分别进行介绍。

5.6.1 一对多映射

一对多映射在设计实体时主要用到下面注解：

- @OneToMany：放在主表中；
- @ManyToOne：放在从表中。

例如，假设一个用户允许有多个通讯方式，则可设计成用户为主表，通讯方式为从表，在从表中设置外键关联至主表。这是典型的一对多映射关系。下面设计实体类 User：

```
@Entity//标识该类为一个实体类
@Table(name = "user")//关联数据库中的 user 表
public class User {
    @Id
    @GeneratedValue(strategy = GenerationType.IDENTITY)
    private Long id;
    @NotNull
    private String name;
    private String address;
    private Date birthday;

    @OneToMany(cascade = CascadeType.ALL, fetch = FetchType.LAZY, mappedBy = "user")
    private List<Phone> phone = new ArrayList<Phone>();
```

```java
    public User() {
    }

    public User(String name, String address, Date birthday) {
        this.name = name;
        this.address = address;
        this.birthday = birthday;
    }

    public User(String name, String address, Date birthday, List<Phone> phone) {
        this.name = name;
        this.address = address;
        this.birthday = birthday;
        this.phone = phone;
    }

    //省略 get 和 set
}
```

在上面代码中，关键是注解：

```java
@OneToMany(cascade = CascadeType.ALL, fetch = FetchType.LAZY, mappedBy = "user")
private List<Phone> phone = new ArrayList<Phone>();
```

表示该实体类为一对多映射，多的一方为 List<Phone>。其中属性 cascade 表示级联操作，取值可分别为：

- CascadeType.PERSIST：级联持久化（保存）操作；
- CascadeType.MERGE：级联更新（合并）操作；
- CascadeType.REFRESH：级联刷新操作，只会查询获取操作；
- CascadeType.REMOVE：级联删除操作；
- CascadeType.ALL：级联以上全部操作。

以上属性设置会影响对该主类中从表操作，比如保存主类表时，从表内容是否自动保存。上面示例 User 类中，phone 上设置为 CascadeType.ALL，表示从表将随主表一起操作，减少人为干预。属性 fetch 表示关联表数据加载方式，取值可分别为：

- FetchType.EAGER：关联数据同时加载；
- FetchType.LAZY：关联数据不加载。

该项设置需依据具体情况，如果从表关联数据较多，或从表太多，一次性加载数据则会产生加载速度慢。如果从表数据量可控，可设置为立即加载关联数据，否则设置为懒加载，在上面设置中，设置为懒加载，则关联数据需要手动加载。

注意：如果该项没有设置，@OneToMany 默认为懒加载，@ManyToOne 默认为立即加载。如果设置为懒加载时，需要将：

```java
@NotFound(action = NotFoundAction.IGNORE)
```

去掉，否则懒加载将失效。

属性 mappedBy 表示关系维护。设置值必须和从表中@ManyToOne 对应字段保持一致,否则将启动出错。

下面是实体类 Phone 示例代码:

```java
@Entity                          //标识该类为一个实体类
@Table(name = "phone")           //关联数据库中的 phone 表
public class Phone {
    @Id
    @GeneratedValue(strategy = GenerationType.IDENTITY)
    @Column(name = "phone_id")
    private Long id;

    @Column(name = "phone_number")
    private  String phoneNumber;

    @Column(name = "note")
    private String notes;

    @ManyToOne(fetch = FetchType.EAGER)
    @JoinColumn(name = "user_id")
    private User user;

    @Column(name = "user_id", insertable = false, updatable = false)
    private Long userId;

    //省略 get 和 set
}
```

在上面代码中,重要设置是:

(1) @ManyToOne:表示该表为从表多对一,指向主表,主表在下面定义,即 User,其中属性 fetch 表示取值方式。

(2) @JoinColumn:表示在从表中关联的字段名称,属性 name 表示字段名称,如果没有该注解,则会默认生成一个关联字段。

(3) @Column:表示该类中字段名和数据表中的字段名对应关系,可包含如下属性:

① name:映射的列名。如:映射 phone 表的 phone_id 列,可以在 id 属性的上面或 getName 方法上面加入;

② unique:表示是否唯一;

③ nullable:表示是否允许为空;

④ length:对应字符型字段,指定字段的最大字符长度;

⑤ insertable:表示是否允许插入;

⑥ updatetable:表示是否允许更新;

⑦ columnDefinition:表示定义建表时创建此列的 DDL;

⑧ secondaryTable:表示从表名。如果此列不建在主表上(默认是主表),该属性定义该列所在从表的名字。

通过以上设置,完成一对多实体设置。下面定义 Dao 类:

```java
public interface UserDao extends JpaRepository<User,Long> {
}
```

和

```
public interface PhoneDao extends JpaRepository<Phone,Long> {
    List<Phone> findByUser_id(Long i);
}
```

以上两个类继承自 JpaRepository，分别实现对 User 和 Phone 的操作，根据接口规范名方法，自定义方法 findByUser_id，用于根据 user_id 操作该用户对应的 phone 信息。下面是具体操作方法：

```
@Service
public class UserService {
    @Autowired
    UserDao userDao;
    @Autowired
    PhoneDao phoneDao;

    public void usesaveAll() throws ParseException {
        //定义两个 Phone 实例
        Phone p1 = new Phone("1381343","beizhu");
        Phone p2 = new Phone("1388876","bygfcj");

        List<Phone> phone = new ArrayList<Phone>();
        phone.add(p1);
        phone.add(p2);
        //定义一个 User 实例
        User user = new User("Tom","Peking",strTodate("2000-2-4"),phone);
        //下面连着 set 非常重要，可以设置 phone 中用户的信息
        p1.setUser(user);
        p2.setUser(user);

        userDao.save(user);                    //同时,级联更新
        //列举指定用户对应 phone 列表
        List<Phone> phonelist = phoneDao.findByUser_id(1L);
        System.out.println("phonelist -> " + phonelist);
    }
    //省略其他
}
```

在以上定义用户服务中，方法 usesaveAll() 中为具体操作方法，如何保存 User 和 Phone 实体，同时，演示如何根据用户 id 查询该用户下所有 phone，并打印输出。以上自定义类中方法的调用示例：

```
userService.usesaveAll();
```

具体代码可参考本节源码。运行以上代码，查看生成创建表 SQL 如下所示：

```
create table phone (
    phone_id bigint generated by default as identity,
    note varchar(255),
    phone_number varchar(255),
```

```
        user_id bigint,
        primary key (phone_id)
    )
    create table user (
        id bigint generated by default as identity,
        address varchar(255),
        birthday timestamp,
        name varchar(255) not null,
        primary key (id)
    )
    alter table phone
        add constraint FKb0niws2cd0doybhib6srpb5hh
        foreign key (user_id)
        references user
```

以上创建语句很清楚表达了两张数据表间一对多关系映射。

5.6.2 一对一映射

表之间一对一映射很好理解,即一张主表中单条记录关联一张从表中的一条记录,比如用户主表单条记录至多只对应用户详细表中一条记录,在数据量大时,能有效提升数据库访问效率。一对一映射主要用到下面注解:

- @OneToOne;
- @JoinColumn。

下面以主表用户基本信息和从表用户详细信息为例,介绍一对一映射关系,设计实体类用户主表 User 代码如下所示:

```
@Entity//标识该类为一个实体类
@Table(name = "user")                                    //关联数据库中的 user 表
public class User {
    @Id
    @GeneratedValue(strategy = GenerationType.IDENTITY)
    private Long id;
    @NotNull
    private String name;
    private String address;
    private Date birthday;
    @Column(name = "userdetail_id", insertable = false, updatable = false)
    private Long userdetailIid;

    @OneToOne(cascade = CascadeType.ALL)
    @JoinColumn(name = "userdetail_id")
    private UserDetail userDetail;

    //省略 get 和 set
}
```

在上面代码中,注解@OneToOne 表示一对一关系,注解@JoinColumn 表示该表中关联从表的外键。注意：在类中字段 userdetailId 上需设置注解 @Column 中属性：insertable = false，updatable = false。

实体类用户详细信息表 UserDetail 代码如下所示：

```
@Entity//标识该类为一个实体类
@Table(name = "userdetail")                              //关联数据库中的 user 表
public class UserDetail {
    @Id
    @GeneratedValue(strategy = GenerationType.IDENTITY)
    private Long id;
    private int age;
    private String workAddress;
    @Lob
    private String notes;

    @OneToOne(cascade = CascadeType.ALL, mappedBy = "userDetail")
    private User user;

    //省略 get 和 set
}
```

上面代码中，用户详细信息表根据需要可能会有更多字段，其中关键是类字段 user 上注解 @OneToOne 设置，mappedBy 表示关系维护，其属性值一定在类 User 中有定义。注解 @Lob 表示该类字段对应数据表字段为 clob 字段。一对一映射中，在获取时，如果没有设置 fetch 方式，默认是关联数据立即加载。

下面是设置接口类 UserDao：

```
public interface UserDao extends JpaRepository<User,Long> {
}
```

一般，如果继承自 JpaRepository，JPA 提供的方法足够用，则可不用自定义其他方法，以提高开发效率。下面是开发服务层：

```
@Service
public class UserService {
    @Autowired
    UserDao userDao;

    public void usesaveAll() throws ParseException {
        //定义 UserDetail
        UserDetail userDetail = new UserDetail(11,"beijing haidian","备注说明");
        //定义 User
        User user = new User("Tom","Peking",strTodate("2000 - 2 - 4"),userDetail);

        userDao.save(user);                              //同时,级联更新

        Optional<User> user2 = userDao.findById(1L);     //同时,级联取出
        System.out.println(user2);

    }
```

```
        //省略其他内容
}
```

上面代码中，主要定义 userDetail 和 user，然后级联保存，完成一对一关系数据表内容的保存。接着在适当地方调用上面定义的 usesaveAll()方法，如下所示：

```
userService.usesaveAll();                              //批量保存
```

运行以上代码，可查看到自动生成 SQL 语句类似如下所示：

```
create table user (
    id bigint generated by default as identity,
    address varchar(255),
    birthday timestamp,
    name varchar(255) not null,
    userdetail_id bigint,
    primary key (id)
)

create table userdetail (
    id bigint generated by default as identity,
    age integer not null,
    notes clob,
    work_address varchar(255),
    primary key (id)
)

alter table user
    add constraint FKb1fqlot3t18igd3f46k1sjkfy
    foreign key (userdetail_id)
    references userdetail
```

上面 SQL 语句与前面实体类设计相对应，表示主从表一对一关系 user 和 userdetail。

5.6.3　多对多映射

两张数据表存在多对多映射，比如用户表和角色表之间，一个用户对应多个角色，而一个角色也同时对应多个用户，即表示多对多关系，在设置中，用到如下注解：

- @ManyToMany；
- @JoinTable。

下面以用户表和角色表为例，介绍如何表示多对多映射：

```
@Entity//标识该类为一个实体类
@Table(name = "user")                                 //关联数据库中的 user 表
public class User {
    @Id
    @GeneratedValue(strategy = GenerationType.IDENTITY)
    private Long id;
    @NotNull
    private String name;
```

```java
    private String address;
    private Date birthday;

    @ManyToMany (cascade = CascadeType.ALL)
    @JoinTable (
        name = "user_roles",                                    //关联表名
        inverseJoinColumns = @JoinColumn (name = "user_id"),    //被维护端外键
        joinColumns = @JoinColumn (name = "role_id" ))          //维护端外键被维护端注解

    private List<Roles> roles;
    //省略 get 和 set
}
```

上面代码中，在定义 roles 上的注解 @ManyToMany，表示多对多关系，注解 @JoinTable 是关键，其中属性有：

- Name：表示该连接表的表名；
- JoinColumns：该属性值可接受多个@JoinColumn，用于设置连接表中外键列的信息；
- inverseJoinColumns：该属性值可接受多个@JoinColumn，用于设置连接表中外键列的信息；
- targetEntity：该属性指定关联实体的类名。在默认情况下，Hibernate 将通过反射来判断关联实体的类名；
- catalog：设置将该连接表放入指定的 catalog 中。如果没有指定该属性，连接表将放入默认的 catalog；
- schema：设置将该连接表放入指定的 schema 中。如果没有指定该属性，连接表将放入默认的 schema；
- uniqueConstraints：该属性用于为连接表增加唯一约束；
- indexes：该属性值为@Index 注解数组，用于为该连接表定义多个索引。

下面是实体类 Roles 的定义：

```java
@Entity//标识该类为一个实体类
@Table(name = "role")                           //关联数据库中的 role 表
public class Roles {
    @Id
    @GeneratedValue(strategy = GenerationType.IDENTITY)
    @Column(name = "role_id")
    private int id;

    @Column(name = "role_name")
    private String roleName;

    @Column(name = "note")
    private String notes;

    @ManyToMany(cascade = CascadeType.REFRESH,
            mappedBy = "roles",                 //通过维护端的属性关联
            fetch = FetchType.LAZY)
```

```java
    private List<User> user;
    //省略 get 和 set
}
```

下面示例如何关联使用以上两个实例：

```java
@Service
public class UserService {
    @Autowired
    UserDao userDao;
    @Autowired
    RolesDao rolesDao;

    public void usesaveAll() throws ParseException {
        Roles r1 = new Roles("juese1","notes1");
        Roles r2 = new Roles("juese2","notes2");

        List<Roles> roles = new ArrayList<Roles>();
        roles.add(r1);
        roles.add(r2);

        User user = new User("Tom","Peking",strTodate("2000-2-4"),roles);

        userDao.save(user);                    //同时,级联更新

        Optional<User> user2 = userDao.findById(1L);
        System.out.println(user2);

        List<Roles> listR = rolesDao.findByUser_Name("Tom");
        System.out.println(listR);
    }
    //省略其他内容
}
```

以上代码中,在方法 usesaveAll() 中先定义多个 Roles 实例,接着定义 User 实例,调用通用保存方法,进行级联保存,最后查看保存信息。其中,rolesDao.findByUser_Name 是用户自定义方法,即如何通过用户的姓名查询其对应的所有角色,该方法定义放在类 RolesDao 中,如下所示：

```java
public interface RolesDao extends JpaRepository<Roles,Long> {
    //自动生成 JPQL user.name:查询指定用户下的所有角色
    List<Roles> findByUser_Name(String userName);
}
```

注意：多对多映射中,默认 fetch 方式是懒加载。

通过上面代码,自动生成 SQL 语句类似如下所示：

```sql
create table role (
    role_id integer generated by default as identity,
    note varchar(255),
    role_name varchar(255),
```

```
        primary key (role_id)
)

create table user (
        id bigint generated by default as identity,
        address varchar(255),
        birthday timestamp,
        name varchar(255) not null,
        primary key (id)
)

create table user_roles (
        role_id bigint not null,
        user_id integer not null
)

alter table user_roles
        add constraint FKili9447fgjyx1siebjfafpoi2
        foreign key (user_id)
        references role

alter table user_roles
        add constraint FKnljwyse14j5ewmp5a08w36o59
        foreign key (role_id)
        references user
```

通过以上 SQL 语句可知，对于多对多映射关系，将生成中间映射表，以关联两张数据表。主要代码请查看本节源码。

5.7 动态查询

Spring Data JPA 围绕查询，提供多种方法，便于开发人员选用。但实际上，还不一定满足用户需求，比如动态查询，更为复杂的查询等。那么，Spring Data JPA 支持 JPA2.0 的 Criteria 查询，Criteria 查询是一种类型安全和更面向对象的查询，其提供一个能够在运行时动态地构建查询的机制。

使用 Criteria 方式需要继承 JpaSpecificationExecutor 接口，该接口是用来负责查询，提供如下一些方法：

- T findOne(Specification<T> var1);
- List<T> findAll(Specification<T> var1);
- Page<T> findAll(Specification<T> var1, Pageable var2);
- List<T> findAll(Specification<T> var1, Sort var2);
- long count(Specification<T> var1);

在具体开发中，可通过实现 Specification 中的 toPredicate() 方法来定义动态查询，并通过 CriteriaBuilder 来创建查询条件，toPredicate() 方法定义如下：

```
Predicate toPredicate(Root<T> root, CriteriaQuery<?> query, CriteriaBuilder cb)
```

为了理解如何使用，下面通过示例进行介绍。

创建 Dao 层 UserDao，需要继承接口 JpaSpecificationExecutor，代码如下所示：

```
package com.zioer.dao;

import org.springframework.data.jpa.repository.JpaSpecificationExecutor;
import org.springframework.data.repository.CrudRepository;
import com.zioer.model.User;

public interface UserDao
    extends CrudRepository<User, Long>, JpaSpecificationExecutor {
}
```

JpaSpecificationExecutor 不能单独使用，如上代码所示。下面在服务层调用 JpaSpecificationExecutor 接口。创建服务层 UserService，并在其中自动装配 UserDao，如下代码所示：

```
@Service
public class UserService {
    @Autowired
    UserDao userDao;

}
```

在类 UserService 中创建方法：

```
public void findUserByNameAndAddress(String name, String address) {

    List<User> list = userDao.findAll(new Specification<User>() {
        @Override
        public Predicate toPredicate( Root < User > root, CriteriaQuery <?> query,
CriteriaBuilder criteriaBuilder) {
            //构建查询条件并返回
            //root.get("address")表示获取address这个字段名称，like表示执行like查询
            Predicate p1 = criteriaBuilder.equal(root.get("name"), name);
            Predicate p2 = criteriaBuilder.equal(root.get("address"), address);
            //将两个查询条件联合起来之后返回Predicate对象
            return criteriaBuilder.and(p1,p2);
        }
    });

    System.out.println("found : " + list);
}
```

上面示例代码中，进行简单的联合查询，调用 toPredicate() 方法：

- Root<User> root：理解为类似于 JPQL 或 SQL 查询的 FROM 子句，表示一个对 User.class 进行计算的表达式。root.get 可理解为从类中取出相应的字段；
- CriteriaBuilder criteriaBuilder：用于创建 CriteriaQuery 条件方法，包括 equal、

notEqual、gt、ge、lt、le、between 和 like 等。

以上代码创建的 SQL 语句类似如下所示：

```
select
  user0_.id as id1_1_,
  user0_.address as address2_1_,
  user0_.birthday as birthday3_1_,
  user0_.name as name4_1_
from
  user user0_
where
  user0_.name = ?
  and user0_.address = ?
```

在上面 toPredicate() 方法中，query 参数没有被使用，下面演示如何使用该参数：

```java
public void findUserByNameAndAddressUserOrderby(String name, String address) {

    List<User> list = userDao.findAll(new Specification<User>() {
        @Override
        public Predicate toPredicate(Root<User> root, CriteriaQuery<?> query, CriteriaBuilder criteriaBuilder) {
            //构建查询条件并返回
            //root.get("address")表示获取 address 这个字段名称,like 表示执行 like 查询
            Predicate p1 = criteriaBuilder.equal(root.get("name"), name);
            Predicate p2 = criteriaBuilder.equal(root.get("address"), address);

            query.where(criteriaBuilder.and(p1,p2));
            //添加排序的功能
            query.orderBy(criteriaBuilder.desc(root.get("id").as(Long.class)));
            return query.getRestriction();
        }
    });

    System.out.println("found : " + list);
}
```

以上代码中，创建一个带 Order by 功能的 SQL，如果可能，还可以创建更为复杂的 SQL。下面代码创建一个复杂的 SQL：

```java
public void findUserByNameOrAddress() {
    //第一个 Specification 定义了两个 or 的组合
    Specification<User> s1 = new Specification<User>() {
        @Override
         public Predicate toPredicate(Root<User> root, CriteriaQuery<?> query, CriteriaBuilder criteriaBuilder) {
                Predicate p1 = criteriaBuilder.equal(root.get("id"),"1");
                Predicate p2 = criteriaBuilder.equal(root.get("id"),"2");
                return criteriaBuilder.or(p1,p2);
        }
    };
    //第二个 Specification 定义了两个 or 的组合
```

```java
        Specification<User> s2 = new Specification<User>() {
            @Override
            public Predicate toPredicate(Root<User> root, CriteriaQuery<?> criteriaQuery, CriteriaBuilder criteriaBuilder) {
                Predicate p1 = criteriaBuilder.like(root.get("address"),"e%");
                Predicate p2 = criteriaBuilder.like(root.get("name"),"T%");
                //将两个查询条件联合起来之后返回 Predicate 对象
                return criteriaBuilder.or(p1,p2);
            }
        };

        List<User> list = userDao.findAll(Specification.where(s1).and(s2));

        System.out.println("found : " + list);
}
```

以上代码创建的 SQL 语句中的 Where 子句如下所示:

```
where
    (
        user0_.id = 1
        or user0_.id = 2
    )
    and (
        user0_.address like ?
        or user0_.name like ?
    )
```

至此,通过构造两个 Specification 实例,进行联合查询,创建一个复杂的 SQL。为了一个更通用的方法,下面创建一个动态 SQL:

```java
public Page<User> findUserDy(User user, int page, int size) {
    Sort sort = new Sort(Sort.Direction.DESC, "id","name");
    Pageable pageable = PageRequest.of(page, size, sort);

    return userDao.findAll(new Specification<User>() {
        @Override
        public Predicate toPredicate (Root<User> root, CriteriaQuery<?> query, CriteriaBuilder criteriaBuilder) {
            List<Predicate> predicateList = new ArrayList<>();

            if (!StringUtils.isEmpty(user.getId())) {
                predicateList.add(criteriaBuilder.equal(root.get("id"), user.getId()));
            }
            if (!StringUtils.isEmpty(user.getName())) {
                predicateList.add(criteriaBuilder.like(root.get("name"), user.getName()));
            }
            if (null != user.getBirthday()) {
                predicateList.add(criteriaBuilder.greaterThan(root.get("birthday"), user.getBirthday()));
            }
```

```
                Predicate[] predicateArr = new Predicate[predicateList.size()];
                return criteriaBuilder.and(predicateList.toArray(predicateArr));
            }
        }, pageable);
    }
```

以上代码是一个稍微复杂的示例,利用 Sort 和 Pageable,进行排序和分页,在 toPredicate 方法中,通过多个 if 判断,生成一个动态 SQL 语句返回,where 子句类似如下:

```
where
    user0_.name like ?
order by
    user0_.id desc,
    user0_.name desc limit ?
```

提示:where 子句中,会根据传入条件不同而有所区别。下面是一个多表联合查询示例:

```
public List<User> findUserByUserAndRole(String roleId) {
    return userDao.findAll(new Specification<User>() {
        @Override
        public Predicate toPredicate(Root<User> root,
        CriteriaQuery<?> query, CriteriaBuilder criteriaBuilder) {
            //方式1
            ListJoin<User, Roles> roleJoin = root.join(root.getModel().getList("roles",
Roles.class), JoinType.LEFT);

            //方式2:更简洁
            //Join<User, Roles> userJoin = root.join("roles", JoinType.LEFT);

            Predicate predicate = criteriaBuilder.equal(roleJoin.get("id"), roleId);

            return predicate;
        }
    });
}
```

上面代码演示如何通过 Join 方式左连接两实体。

本节介绍 Spring Data JPA 如何灵活地创建查询语句,以满足开发人员更加丰富的查询需要。以上代码和运行效果请查看本节源码。

5.8 简单配置

以上介绍 Spring Data JPA 时,在 pom.xml 文件中加入相关依赖,并没有进行任何配置,便完成案例及代码讲解。实际上,以上介绍基于前面介绍 H2 数据库,在 pom.xml 文件中加入下面依赖:

```
<dependency>
```

```
<groupId>com.h2database</groupId>
<artifactId>h2</artifactId>
</dependency>
```

不作任何配置,默认使用 H2 数据库的内存模式,方便开发人员进行程序设计时调试工作。基于此,通过 Spring Data JPA 提供的@Entity 注解,完成数据表自动创建工作,使得很容易完成 Spring Data JPA 案例分析。

但在实际开发中,根据需要还是有一定的配置工作,比如默认情况下,创建、插入等 SQL 不显示,不利于开发时的调试工作。下面简单介绍关于 Spring Data JPA 的配置。

在配置文件 application.properties 中加入下面配置:

```
spring.jpa.properties.hibernate.show_sql=true
spring.jpa.properties.hibernate.use_sql_comments=true
spring.jpa.properties.hibernate.format_sql=true
```

使得程序在运行时,Console 窗口可以格式化显示运行时 SQL 以相关说明,利于调试工作,当然,最好在系统部署前删掉上面代码,使得系统运行更快、效率更高。加入下面配置:

```
spring.datasource.url=jdbc:h2:mem:h2test;
    DB_CLOSE_DELAY=-1;DB_CLOSE_ON_EXIT=FALSE
spring.datasource.platform=h2
spring.datasource.username=sa
spring.datasource.password=
spring.datasource.driver-class-name=org.h2.Driver
```

以上配置项表示显示配置程序使用数据库为 H2,运行方式为内存方式,可根据需要更改为其他数据库。加入下面配置:

```
spring.jpa.database-platform=org.hibernate.dialect.H2Dialect
spring.jpa.hibernate.ddl-auto=update
```

其中,ddl-auto 取值如下:

- create:每次运行该程序,没有表格会新建表,表内有数据会清空;
- create-drop:每次程序结束的时候会清空表;
- update:每次运行程序,没有表格会新建表,表内有数据不会清空,只会更新;
- validate:运行程序会校验数据与数据库的字段类型是否相同,不同会报错;
- none:不创建表。

以上配置项用于是否采用业务实体方式自动创建数据表,关键点是 ddl-auto 的取值。关于创建数据表还有一种方式是配置 sql 文件,即将创建数据表语句放入某一个 sql 文件中,然后配置系统启动时,自动创建数据表。一种方式是在 resources 目录下创建 schema.sql 文件,不作任何配置,系统启动时,将自动运行 schema.sql 中的语句;另一种方式是创建任意文件名的 sql 文件,可放入工程目录下任意位置,配置文件中加入类似下面配置:

```
spring.datasource.schema=classpath:db/schema.sql
spring.datasource.data=classpath:db/data.sql
```

data.sql 文件中是初始数据语句。当加入以 sql 结尾文件方式创建的数据表后,注意和前面 ddl-auto 配置关系。如果两种方式都进行了配置，则系统运行时，首先运行 sql 文件，然后检测 ddl-auto 方式；如果不想 ddl-auto 干扰，则只需将其配置为 none 即可。

本章小结

　　本章详细讲解了 Spring Data JPA 的各种用法即相关配置，并在每节中加入翔实案例分析，有助于理解。Spring Data JPA 是面向对象操作数据库的一种比较流行的方式。其优点是，开发人员可在不理解 SQL 语句的情况下进行快速开发，操作也不是很复杂。结合本章讲解案例，可对这种方式进行深入理解。

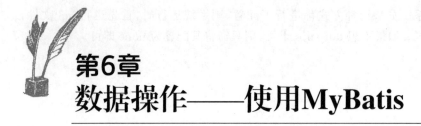

第6章 数据操作——使用MyBatis

本章介绍 Spring Boot 2 中关系数据库操作使用较多的优秀开源持久层框架 MyBatis，其具备可定制化 SQL、存储过程以及高级映射。

6.1 MyBatis 介绍

上一章介绍了基于对象操作方式操作数据库的 Spring Data JPA，但不是所有开发人员都喜欢这样一种方式。部分开发人员还是喜欢直接操作 SQL 语句，但 JDBC 方式又有很多弊端。本章介绍的 MyBatis 就是迎合这一部分开发人员的开源持久层框架。MyBatis 最大优点是避免了几乎所有的 JDBC 代码和手动设置参数以及获取结果集，目前，其支持使用简单的 XML 或注解来配置和映射原生信息，将接口和 Java 的 POJOs（Plain Ordinary Java Object，即普通的 Java 对象）映射到数据库中的记录。

简单理解 MyBatis 功能架构，可分为三层：

- API 接口层：用于提供给外部使用的 API 接口，开发人员通过这些接口来操作数据库。该层一旦接收到调用请求，就调用数据处理层来完成相关数据处理；
- 数据处理层：负责具体的 SQL 查找、SQL 解析、SQL 执行和执行结果映射处理等；
- 基础支撑层：负责基础的功能支撑，包括连接管理、事务管理、配置加载和缓存处理等，主要目的是为其上的数据处理层提供最基础的支撑。

其关系如图 6.1 所示。

MyBatis 具有如下特点：

- 简单易学：没有其他依赖，通过简单配置，很容易入门；
- 灵活：SQL 可与代码分开管理，即单独写在 XML 里，便于统一管理和优化，或可作为注解方式管理；
- 解耦：提供 DAO 层，可做到 SQL 和代码的分离，提高可维护性；

图 6.1 MyBatis 功能框架图

- 支持对象与数据库字段关系映射；
- 支持编写动态 SQL；
- 易于 SQL 调优。

关于 MyBatis 最为强大的部分是支持动态 SQL 的编写，其采用 OGNL（Object-Graph Navigation Language）表达式，使开发人员可以很容易编写出动态 SQL，满足业务需求。在没采用 MyBatis 之前，很多开发人员应该经历过为了拼写正确 SQL，经常忘记或多写某列前后的逗号等，MyBatis 在这方面提供了很好的解决方案。

6.2 快速入门

正由于 MyBatis 如此强大，很受开发人员喜爱。Spring Boot 2 对 MyBatis 提供了很好支持。本节将通过一个简单案例，介绍 Spring Boot 2 中如何快速通过 MyBatis 操作数据库。

创建一个 Spring Boot 2 工程，编辑 pom.xml 文件，加入 MyBatis 相关依赖，该文件中依赖部分代码如下所示：

```xml
<dependencies>
    <dependency>
        <groupId>org.springframework.boot</groupId>
        <artifactId>spring-boot-starter</artifactId>
    </dependency>
    <dependency>
        <groupId>org.mybatis.spring.boot</groupId>
        <artifactId>mybatis-spring-boot-starter</artifactId>
        <version>2.1.0</version>
    </dependency>
    <dependency>
        <groupId>com.h2database</groupId>
        <artifactId>h2</artifactId>
        <scope>runtime</scope>
    </dependency>
    <dependency>
        <groupId>org.springframework.boot</groupId>
        <artifactId>spring-boot-starter-test</artifactId>
        <scope>test</scope>
    </dependency>
</dependencies>
```

如上代码所示，mybatis-spring-boot-starter 依赖是和 MyBatis 相关，该依赖将加入 MyBatis 的 jar 支持，h2 依赖表示使用 H2 数据库，本章还是使用 H2 数据库，保存该文件。编辑配置文件 application.properties，加入下面内容：

```
spring.datasource.schema=classpath:db/schema.sql
spring.datasource.data=classpath:db/data.sql
```

```
# MyBatis
mybatis.mapper-locations=classpath:mapper/*.xml
```

上面代码中，schema 和 data 中加入创建数据表的 DDL 和 DML 相关语句，目的是系统启动时，自动在内存中创建数据表和初始化数据。第二部分配置和 MyBatis 有关，定义 mapper 层，即 SQL 语句相关 xml 文件放置位置。schema.sql 文件只定义了一张数据表，如下所示：

```
create table user (
    id bigint generated by default as identity,
    name varchar(25) not null,
    address varchar(255),
    sex varchar(10),
    primary key (id)
);
```

data.sql 文件是初始数据插入语句，如下所示：

```
INSERT INTO 'user' VALUES ('1', 'Tom','Peking', '男');
INSERT INTO 'user' VALUES ('2', 'Kitty','Shanghai', '女');
```

下面是定义模型层代码：

```
package com.zioer.model;

public class User {
    private Long id;
    private String name;
    private String address;
    private String sex;

    //省略下面 get 和 set
}
```

上面模型比较简单，和同名数据表对应。下面定义 mapper 层：

```
package com.zioer.mapper;
@Mapper
public interface UserMapper{
    //查询全部数据
    List<User> selectAll();
}
```

mapper 层定义的是接口，下面定义 mapper 层相关 XML 文件 UserMapper.xml：

```
<?xml version="1.0" encoding="UTF-8"?>
<!DOCTYPE mapper PUBLIC "-//mybatis.org//DTD Mapper 3.0//EN"
    "http://mybatis.org/dtd/mybatis-3-mapper.dtd">

<!-- 当 Mapper 接口和 XML 文件关联的时候，namespace 的值就需要配置成接口的全限定名称 -->
<mapper namespace="com.zioer.mapper.UserMapper">
    <select id="selectAll" resultType="com.zioer.model.User">
```

```
        select id,    name , address , sex from user
    </select>
</mapper>
```

上面代码实际上就是 XML 文件格式,元素 mapper 中属性 namespace 值对应前面定义的接口 UserMapper。在这里,只定义了简单查询语句,元素 select 中属性 id 的值需和前面 UserMapper 接口中的一个方法名相同。至此,mapper 层定义完成。定义服务层接口:

```
package com.zioer.service;

public interface UserService {
    List<User> getUserList();
}
```

上面接口中,只有一个方法待实现,下面定义实现类:

```
package com.zioer.service.impl;

@Service
public class UserServiceImpl implements UserService{
    @Autowired
    private UserMapper userMapper;

    @Override
    public List<User> getUserList() {
        return userMapper.selectAll();
    }
}
```

在上面实现类中,继承自接口 UserService,同时在该类上加注解@Service,以注册该类为服务,并在其中实现方法 getUserList(),调用 mapper 层中 selectAll(),返回 select 全部记录。至此,MyBatis 简单相关内容代码已完成,下面只需要在相关地方调用上面实现类中的 getUserList()方法即可。简单起见,直接在该工程中的 Application 类中,加入下面方法调用:

```
@Override
public void run(String... args) throws Exception {
    List<User> list = userService.getUserList();
    for(User user:list) {
        System.out.println(user);
    }
}
```

保存后,执行该工程,在 Console 窗口中将列举出数据表中全部记录,如图 6.2 所示。

```
2019-08-04 16:01:10.845  INFO 35632 --- [
{id=1, name='Tom', address='Peking', sex='男'}
{id=2, name='Kitty', address='Shanghai', sex='女'}
{id=3, name='Jim', address='Guangzhou', sex='男'}
{id=4, name='Yammi', address='Chengdu', sex='男'}
```

图 6.2 运行结果示例

该工程目录如图 6.3 所示。

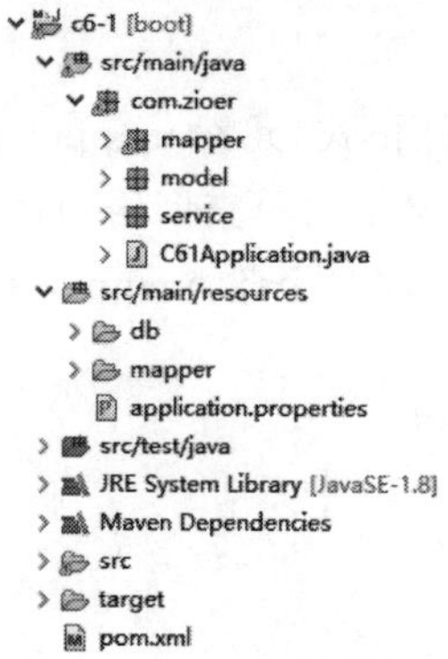

```
v 📂 c6-1 [boot]
  v 📂 src/main/java
    v 📦 com.zioer
      > 📦 mapper
      > 📦 model
      > 📦 service
        📄 C61Application.java
  v 📂 src/main/resources
    > 📂 db
    > 📂 mapper
      📄 application.properties
  > 📂 src/test/java
  > 📚 JRE System Library [JavaSE-1.8]
  > 📚 Maven Dependencies
  > 📂 src
  > 📂 target
    📄 pom.xml
```

图 6.3　示例工程目录

本节示例比较简单，关键是在 pom.xml 中加入 MyBatis 相关依赖，SQL 语句和 Java 代码分开管理，代码分层为 model、mapper 和 service 层，SQL 语句集中在 mapper 层的 xml 文件中，层次结构清晰。具体代码和运行效果见本节源码。

6.3　MyBatis 基本元素

通过上节示例，很容易理解 MyBatis 就是一种 SQL 语句集中管理方式，即将 SQL 语句全部放入 xml 文件中。为了更好理解这种形式，本节介绍 MyBatis 的基本组成元素。在 xml 文件中，基本表现形式如下：

```
< mapper namespace = " … ">
    …
</mapper >
```

该 xml 文件根元素 mapper，属性 namespace 指定 mapper 接口位置，使用全限定名。Namespace 在绑定接口后，可以不用再写接口实现类，MyBatis 会通过该绑定自动找到对应要执行的 SQL 语句。在 mapper 元素中，包含以下几个重要元素，以实现对数据表的查询、新增、更新和删除操作。

(1) select 元素

该元素用于查询数据，查询语句是所有业务系统中的重要操作。MyBatis 做的就是如何将查询结果进行映射，并返回。前面示例：

```
< select id = "selectAll" resultType = "com.zioer.model.User">
    select id, name, address, sex from user
</select >
```

其中，属性 id 必须在整个 xml 文件中具有唯一值，其和接口文件中的方法名称一致。属性 resultType 表示结果映射，在这里使用了全限定名。

在 select 元素中，可接受用户传递参数进行查询，例如：

```xml
<select id="selectById" parameterType="long" resultType="com.zioer.model.User">
    select id, name, address, sex from user where id = #{id}
</select>
```

以上代码用于查询指定条件的记录。属性 parameterType 表示传入这条语句的参数类的完全限定名或别名。

在上面 SQL 语句中使用 #{} 占位符，#{} 主要用于字符串安全替换。

（2）insert 元素

该元素主要用于在数据表中插入记录。示例代码如下所示：

```xml
<insert id="insertUser" parameterType="com.zioer.model.User">
    insert into user (name,address,sex) values (#{name},#{address},#{sex})
</insert>
```

以上代码中，insert 元素的属性 id 值需要具有唯一性，parameterType 表示传入参数，在这里使用了全限定名。

（3）update 元素

该元素和 insert 元素类似，但用于数据的更新操作，示例代码如下所示：

```xml
<update id="updateUser" parameterType="com.zioer.model.User">
    update user set name = #{name},address = #{address},sex = #{sex} where id = #{id}
</update>
```

提示：上面代码中，SQL 语句中的 id 为数据表关键字，where 条件表示通过关键字 id 修改记录。

（4）delete 元素

该元素用于数据表的删除记录操作，示例代码如下所示：

```xml
<delete id="deleteUser" parameterType="long">
    delete user where id = #{id}
</delete>
```

上面代码只需要传入待删除记录的 id 即可。

以上几个元素是 MyBatis 的基本元素，通过这几个元素便可实现数据表的几个基本操作。MyBatis 在 SQL 规整方面做了很大努力，即将 SQL 语句全部集合放入 xml 文件中，暴露简单接口，以实现数据表的操作。以上代码和运行效果见本节源码。

6.4　#{} 与 ${}

MyBatis 提供了两个重要的字符串替换符号，即 #{} 与 ${}。#{} 在上一节中已经有所接触，主要用于字符串安全替换，类似于 Java 中书写预处理语句中符号？一样，示例代码如下：

```java
String selectUser = "select * from user where id = ?";
PreparedStatement ps = conn.prepareStatement(selectUser);
ps.setInt(1,id);
```

以上方式在Java中书写时很常见,可以将一些非安全字符进行转义,防止SQL注入等错误。#{}是推荐的一种方式,正如上节介绍的几个示例一样。

但不是任何时候都需要将传入的数值转义,比如传入的是数据表名、字段名等,此时就需要非转义方式替换字符,示例SQL代码如下所示:

```
select id, name, address, sex from user order by ${column} ${direction}
```

在上面代码中,传入column和direction都没经过转义进行替换,比如column值为name,direction值为desc,则生成的SQL为:

```
select id, name, address, sex from user order by name desc
```

非转义替换字符可以用来生成动态的元数据,包括表名、字段名等,是非常有用的。但是,需要在Java代码中控制传入字符串是安全的。

6.5 结果映射

在系统开发中,使用最多的是查询操作,那就涉及结果映射。MyBatis提供了多种结果映射方式,包括领域模型映射、Map映射和resultMap方式。

在本章前面节中,已经涉及查询操作的结果映射,即在select元素中使用属性resultType,其类型为com.zioer.model.User,是开发人员自定义的全限定名称,部分代码如下所示:

```
public class User {
    private Long id;
    private String name;
    private String address;
    private String sex;
    ...
}
```

以上为开发人员自定义领域模型User,注意其属性和数据表中字段对应。如果数据表和用户定义领域模型不一致时,此时需要在select时作相应处理。比如user表定义如下:

```
create table user (
    user_id bigint generated by default as identity,
    user_name varchar(25) not null,
    user_address varchar(255),
    user_sex varchar(10),
    primary key (id)
);
```

以上数据表定义和之前领域模型不一致,则在select查询时,可以用列的别名来进行匹配,如下所示:

```
<select id="selectById" parameterType="long" resultType="com.zioer.model.User">
    select user_id as id,
```

```
    user_name as name,
    user_address as address,
    user_sex as sex
    from user where id = #{id}
</select>
```

通过以上字段名别名方式,实现数据表字段名和领域模型匹配。

第二种结果映射方式是 Map 方式,这种方式比较自由,开发人员可以不事先定义业务模型,示例代码如下所示:

```
<select id = "selectById" parameterType = "long" resultType = "map">
    select user_id,
    user_name ,
    user_address ,
    user_sex
    from user
    where user_id = #{id}
</select>
```

以上代码返回结果类型为 map,MyBatis 自动将结果返回为键值对形式,即可采用 HashMap 方式进行处理,在 Java 中可以采用下列形式获取其中值:

```
Map map = userService.getUserById(2L);
String username = map.get("USER_NAME").toString();
```

采用 Map 形式,可以抛弃自定义业务领域模型,自由度更高,缺点是没有模型,代码和数据表耦合度更高。

第三种结果映射方式是 resultMap,这种方式灵活度更高,能更灵活处理开发人员需求,比如数据表中列名和模型中属性名不一致问题,而不需要过多配置。在 xml 文件中 resultMap 元素示例代码如下:

```
<resultMap id = "userResultMap" type = "com.zioer.model.User">
    <id property = "id" column = "user_id" />
    <result property = "name" column = "user_name"/>
    <result property = "address" column = "user_address"/>
    <result property = "sex" column = "user_sex"/>
</resultMap>
```

在上面代码中,元素 id 和 result 是最基本映射关系,用于表示数据表中列名和模型中属性名对应关系,属性 column 表示数据表中列名,属性 property 表示模型中属性名,则 select 写法如下所示:

```
<select id = "selectUseResultMap" resultMap = "userResultMap">
    select
    user_id ,
    user_name,
    user_address,
    user_sex
    from user
</select>
```

在上面代码中,元素 select 中属性 resultMap 值为前面创建元素 resultMap 的 id 值。

以上介绍的三种 select 结果映射方式各有特点,在实际开发中可以根据实际情况灵活运用。本节介绍的示例详见本节源码。

6.6 注解方式

目前,MyBatis 最新版支持注解配置方式。注解方式提供如下几个关键注解词:
- @Select:查询操作;
- @Insert:插入操作;
- @Update:更新操作;
- @Delete:删除操作。

注解方式不再需要 xml 文件,只需要将注解插入到 mapper 层即可,示例代码如下:

```
@Mapper
public interface UserMapper{
    /**
     * 查询全部数据
     */
    @Select("select user_id as id, user_name as name, user_address as address, user_sex as sex from user")
    List<User> selectAll();
}
```

在上面代码中,自定义接口 UserMapper 中方法 selectAll() 上加入注解 @Select,即可实现查询操作,效果和 xml 中元素 select 一样。同样,插入操作示例如下:

```
@Insert("insert into user values (#{id}, #{name}, #{address}, #{sex})")
void insertUser(User user);
```

上面代码中,只需要在方法 insertUser() 上加入注解 @Insert,即实现插入数据操作。更新操作示例代码如下所示:

```
@Update("update user set user_name = #{name}, user_address = #{address}, user_sex = #{sex} where user_id = #{id}")
void updateUser(User user);
```

删除操作示例代码如下所示:

```
@Delete("delete from user where user_id = #{id}")
void deleteUser(@Param("id")long id);
```

在上面代码中,使用了一个新的注解:

@Param:表示一个入参。

通过以上几个示例,使用几个关键注解便可实现数据的基本操作,在一定程度上简化了操作。下面示例代码是采用注解方式实现 xml 文件中的 resultMap 方式:

```
@Select("select * from user")
@Results(
    id = "userResultId",
    value = {
        @Result(column = "user_id",property = "id"),
        @Result(column = "user_name",property = "name"),
        @Result(column = "user_address",property = "address"),
        @Result(column = "user_sex",property = "sex")
    }
)
List<User> selectUseResultMap();
```

以上注解实现了数据表中字段与模型中属性对应方法，简化 SQL 书写，其中几个重要注解如下：

- @Results：表示结果集合；
- @Result：@Results 中子结果。

用到如下属性：

① id：表示该结果结合唯一名称，以便在其他方法中调用；

② column：表示数据表中列名；

③ property：表示模型中属性名，即列名和类中的属性名之间的映射关系。

同样，可以使用注解方式将查询结果返回为 Map，即键值对形式，便于在程序中直接通过查询键的方式得到值，示例代码如下所示：

```
@MapKey("USER_ID")
@Select("select * from user where user_id = #{id}")
Map selectByid(@Param("id")long id);
```

上面代码中，用到的注解@MapKey 表示键值对，其中的值为返回记录中可作为唯一值的字段名，在这里为 USER_ID，大写很重要，默认情况下，返回的键名称默认为大写。

Map 为方法的返回类型，可简单书写为 Map。比如，查询 id 为 2 的记录结果输出如下：

{2 = {USER_SEX = 男, USER_ID = 2, USER_NAME = TianTian, USER_ADDRESS = Zhenzhou}}

以上结果的返回方式和上节介绍返回 Map 形式有所区别，但在程序中同样很好理解和处理。注解@MapKey 还可以映射到模型上，示例代码如下：

```
@MapKey("id")
@ResultMap("userResultId")
@Select("select * from user where user_id = #{id}")
Map selectByid(@Param("id")long id);
```

以上代码中，用到的注解@ResultMap 是用来引用之前结果集合@Results 的 id 值，在这里值为之前定义的结果集合值 userResultId。同时，注意注解@MapKey 的取值变化。

由以上示例可知，注解方式和 xml 方式能实现同样效果。上面介绍的示例详见本节源码。

6.7 动态 SQL

MyBatis 的强大之处在于能根据条件生成动态 SQL。下面介绍在 MyBatis 中如何生成动态 SQL，减轻开发工作量。采用 xml 文件时，提供了如下几个元素：
- if：用于条件判断；
- choose（when，otherwise）：类似于 Java 中 choose 条件判断，只取其中一值；
- foreach：用于集合便利，比如用在 SQL 的 in 中。

在开发中，查询通常不会像前面所介绍那么简单，比如，根据某条件查询指定记录，那么条件判断就很重要，示例代码如下所示：

```xml
<select id="selectUseResultMap" parameterType="map" resultMap="userResultMap">
    select * from user where 1=1
    <if test="name != null">
        and user_name = #{name}
    </if>
</select>
```

在上面代码中，在 select 元素中，parameterType 值为 map，即传入参数方式采用键值对方式，这种传参方式很重要，其灵活性更强。在该元素内容中，加入条件判断 if，用于判断是否有传入值 name，如果值不为 null，则加入条件判断中的内容。这种方式类似于各种编程语言中的 if 判断，很容易理解。但是，为了加入条件，而在 SQL 中加入了"where 1=1"，显得不够优雅。MyBatis 提供了一种更灵活的方式处理这种情况，即加入元素 where。

将上面语句更改如下：

```xml
<select id="selectUseResultMap1" parameterType="map" resultMap="userResultMap">
    select * from user
    <where>
        <if test="name != null">
            and user_name = #{name}
        </if>
        <if test="sex != null">
            and user_sex = #{sex}
        </if>
    </where>
</select>
```

上面代码中，少了 where 关键字，SQL 更加优雅。元素 where 在这里有两个作用，一是如果元素 where 中有内容时，便自动加上 where 关键字，否则不会加上；其次，如果判断其中的内容前面包含 and 或者 or 时，将自动删除。正如上面示例，加入了两个并行判断 if，其中每个内容前面都有 and。这些问题，MyBatis 都自动处理。实际上，元素 if 可以嵌套组成更复杂的条件判断。

有时，条件判断是互斥的，即只需要其中一个成立就行，此时，使用元素 choose 更加方便，示例代码如下：

```xml
<select id="selectUseResultMap" parameterType="map" resultMap="userResultMap">
    select * from user
    <where>
        <choose>
            <when test="name != null">
                and user_name = #{name}
            </when>
            <when test="sex != null">
                and user_sex = #{sex}
            </when>
            <otherwise>
                and user_id = 1
            </otherwise>
        </choose>
    </where>
</select>
```

上面代码中,元素 choose 表示条件选择,同时用到下面元素:

- 元素 when:判断当前条件是否成立,如果成立,则加入其中内容,choose 结束,否则继续判断其他条件;
- 元素 otherwise:当所有元素 when 都不成立时,加入其中内容。加入该元素不是必需的,但加入该元素是个很好的编程习惯。

在开发中,拼凑 SQL 时,可能会用到 IN 关键字,MyBatis 提供了这样一个方法,示例代码如下:

```xml
<select id="selectUseResultMapByList" parameterType="list" resultMap="userResultMap">
select * from user where user_id in
    <foreach item="item" index="index" collection="list"
        open="(" separator="," close=")">
            #{item}
    </foreach>
</select>
```

以上代码中,使用了元素 foreach,将传入参数 list 自动拼凑为 in 值,并以逗号隔开。通过以上几个简单示例,可知 MyBatis 创建动态语言的强大之处,除了以上几个元素之外,MyBatis 还提供了其他有用元素提高开发效率,如下所示:

- 元素 set:应用到更新 SQL 中,自动加入关键字 set,并确定是否自动去除内容中最后的逗号;
- 元素 bind:创建一个变量,并将其应用到上下文;
- 元素 trim:可自定义加入前缀、后缀,或去除指定前缀或后缀值。

限于篇幅,在此不作详细介绍。在使用注解方式时,同样可以实现动态 SQL,示例如下:

```java
@Update({"<script>",
    "update user",
    "   <set>",
```

```
            "   <if test = 'name != null'> user_name = #{name},</if>",
            "   <if test = 'address != null'> user_address = #{address},</if>",
            "   <if test = 'sex != null'> user_sex = #{sex},</if>",
            "   </set>",
            "where user_id = #{id}",
            "</script>"})
    void updateUser(User user);
```

上面代码中,在注解@Update中,使用了<script></script>,这很重要,目前支持写法就如此,可能相对复杂,但同样可以实现动态SQL。

上面介绍的示例详见本节源码。通过源码,了解一个重要知识点,xml 文件和注解方式可同时使用。建议简单 SQL 可直接使用注解方式;复杂 SQL,比如动态 SQL 等,可采用 xml 文件方式。

6.8 几个重要配置

在 Spring Boot 2 中,使用 MyBatis 很方便,开箱即用,前面介绍中,只需在配置文件中加入下面内容:

```
mybatis.mapper-locations = classpath:mapper/*.xml
```

即用于系统启动时扫描 xml 位置,但如果程序中全部采用注解方式,上面的配置都可以省略。对于初学者,快速入门 MyBatis 有很大帮助。但为了更好地利用 MyBatis 特点和发挥其性能,需要了解关于 MyBatis 的一些配置,比如类型别名,可适当减少代码量等。在配置文件中加入下面配置内容:

```
#配置类型别名所在位置
mybatis.type-aliases-package = com.zioer.model
#输出SQL语句,在开发和测试阶段很有用,查看生成的SQL语句
mybatis.configuration.log-impl = org.apache.ibatis.logging.stdout.StdOutImpl
```

以上配置方法是直接在项目的配置文件 application.properties 中书写,即如果配置类型别名后,可在 xml 文件中简化类型的书写,示例代码如下:

```
<resultMap id = "userResultMap" type = "User">
    //省略中间内容
</resultMap>
```

以上代码中 type 的值可简写为 User,而不必写全限定名称 com.zioer.model.User。

另一种书写配置方式是将 MyBatis 相关配置单独写在 xml 文件中,提高其维护性,例如 MyBatis 配置文件为 MyBatisConfig.xml,在 application.properties 加入下面内容以指定 MyBatis 的配置文件所在位置:

```
mybatis.config-location = MyBatisConfig.xml
```

MyBatisConfig.xml 中内容示例如下:

```xml
<?xml version="1.0" encoding="UTF-8"?>
<!DOCTYPE configuration
        PUBLIC "-//mybatis.org//DTD Config 3.0//EN"
        "http://mybatis.org/dtd/mybatis-3-config.dtd">
<configuration>
    <settings>
        <setting name="cacheEnabled" value="true"/>
        <setting name="lazyLoadingEnabled" value="true"/>
        <setting name="multipleResultSetsEnabled" value="true"/>
        <setting name="useColumnLabel" value="true"/>
        <setting name="useGeneratedKeys" value="false"/>
        <setting name="autoMappingBehavior" value="PARTIAL"/>
        <setting name="autoMappingUnknownColumnBehavior" value="WARNING"/>
        <setting name="defaultExecutorType" value="SIMPLE"/>
        <setting name="defaultStatementTimeout" value="25"/>
        <setting name="defaultFetchSize" value="100"/>
        <setting name="safeRowBoundsEnabled" value="false"/>
        <setting name="mapUnderscoreToCamelCase" value="false"/>
        <setting name="localCacheScope" value="SESSION"/>
        <setting name="jdbcTypeForNull" value="OTHER"/>
        <setting name="lazyLoadTriggerMethods" value="equals,clone,
            hashCode,toString"/>
        <setting name="logImpl" value="STDOUT_LOGGING" />
    </settings>
    <typeAliases>
        <package name="com.zioer.model"/>
    </typeAliases>
</configuration>
```

上面代码涉及内容比较全，但需要注意的是里面元素顺序不能前后颠倒，比如元素 setting 不能放在元素 mappers 之后，同时，在系统的配置文件 application.properties 中不能出现 mybatis.configuration 开头的配置，否则都将报错。下面介绍元素 setting 中的内容：

（1）cacheEnabled：设置缓存全局开关。

（2）lazyLoadingEnabled：设置延迟加载全局开关。

（3）multipleResultSetsEnabled：设置允许单一语句返回多个结果集，需要兼容驱动。

（4）useColumnLabel：设置使用列标签代替类名。

（5）useGeneratedKeys：设置允许 JDBC 使用数据库自增主键。

（6）autoMappingBehavior：设置指定自动映射到字段的规则，选项有：

① NONE：取消自动映射；

② PARTIAL：只映射没有定义嵌套结果集映射的结果集；

③ FULL：自动映射任何结果集。

（7）autoMappingUnknownColumnBehavior：设置指定当自动映射碰到未知列的处理规则，具体选项有：

① NONE：不做任何处理；

② WARNING：输入警告日志；

③ FAILING：抛出 SqlSessionException 异常。

（8）defaultExecutorType：设置配置默认执行器，具体选项有：

① SIMPLE：普通执行器；

② REUSE：重用预处理语句；

③ BATCH：重用语句并执行批量更新。

（9）defaultStatementTimeout：设置驱动等待数据库响应的超时时间，注意事项有：

① 该设置项默认没有值；

② 值的范围是任意正整数；

③ 值的单位是秒。

（10）defaultFetchSize：为驱动的结果集获取数量设置值，注意事项有：

① 该设置项默认没有值；

② 值的范围是任意正整数；

③ 可在具体查询中通过 fetchSize 覆盖该设置项。

（11）safeRowBoundsEnabled：设置允许在嵌套语句中使用分页 RowBounds。

（12）mapUnderscoreToCamelCase：设置开启驼峰命令规则自动转换功能。

（13）localCacheScope：设置利用本地缓存机制防止循环引用和加速重复嵌套查询，具体选项有：

② SESSION：缓存一个会话中执行的所有查询；

② STATEMENT：本地会员只用在语句执行中，对相同 SqlSession 的不同调用不会共享数据。

（14）jdbcTypeForNull：设置当没有为参数提供特定的 JDBC 类型时，为空值指定 JDBC 类型，具体选项有：

① OTHER：一般类型；

② NULL：空值；

③ VARCHAR：字符串。

（15）lazyLoadTriggerMethods：设置指定某个对象的方法触发一次延迟加载，多个方法名称通过逗号划分。

（16）logImpl：设置日志的具体实现方式，注意事项有：

① 该设置项没有默认值；

② 值可以是 slf4j/log4j/log4j2/jdk_logging/commons_logging/stdout_logging/no_loggging；

③ 未指定值的时候会在上述支持列表中自动查找。

以上配置及运行见本节详细代码。

本章小结

本章介绍如何在 Spring Boot 2 中使用 MyBatis。通过本章可掌握如何在基于 Spring Boot 2 的项目中加入 MyBatis 依赖，以及调用 MyBatis 所提供方法。尽管本章介绍是基于

H2 关系数据库，实际上基于关系数据库的使用方法基本通用。同时，本章介绍如何使用 MyBatis，涉及常用方法和使用技巧，介绍了两种方式，即 xml 和注解。两种方式各有优缺点，主要看开发人员习惯以及所在开发项目小组约定而定。通过本章介绍，MyBatis 实际上已成为目前较流行的关系数据库应用系统开发中间件，主要在于其学习复杂度低，后期维护和优化简单等特点。本章介绍仅是抛砖引玉，实践是最好的老师，结合本章案例代码分析，希望使开发人员更加容易理解和快速掌握 MyBatis 的使用。

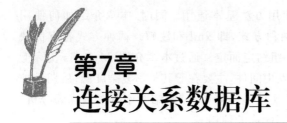

第7章 连接关系数据库

本章介绍 Spring Boot 2 中如何连接多种关系数据库方法。我们知道,连接不同关系数据库都有不同的驱动,本章重点介绍如何通过 Maven 配置文件和简单的连接字符串,连接当前几个主流的数据库,最后,介绍一种在同一个应用中连接多关系数据库的方法,连接多关系数据库存在于很多实际应用中。

7.1 简单介绍

前面用了多章介绍了 Spring Boot 2 操作关系数据库方法,为了简便,全部基于 H2 数据库。当使用 H2、HSQL 和 Derby 等内存数据库时,其优点是:不需要提供任何连接 URL 等,只需加入相应的驱动 Jar 包,Spring Boot 2 便自动发现和装配。

H2 数据库可用于开发阶段和测试阶段,特别是不熟悉各种连接方法前,可快速学习和掌握 Spring Boot 2 操作关系数据库的各种方法和特性,以及进行业务系统的快速开发。但不建议 H2 用于实际生产环境,特别是对各种数据库性能有严格要求环境下。

得益于 Spring 的分层理念,连接数据库层和业务逻辑层可降低耦合度,连接层只关心和数据库连接和传递各种 SQL 语句,业务逻辑层只关心业务逻辑处理。在前面几章中,介绍了 Spring Boot 2 处理和关系数据库的各种方法,并没有涉及 H2 数据库的特性,因为这不重要。

本章将专注于在 Spring Boot 2 中和其他较流行的关系数据库进行连接的配置方法。

7.2 连接 MySQL 数据库

在当前互联网各类网站等数据库应用系统中,MySQL 数据库占相当大比例,这是由于 MySQL 是一种关系数据库管理系统,其具有支持多种操作系统、标准 SQL 查询、提供多种数据库连接以及支持集群等特点,具有简便性和灵活性等特点,深受广大开发人员的喜爱并被广泛应用。

在 MySQL 官网下载 MySQL 最新稳定版,官网如下:

https://www.mysql.com/

根据需要,下载对应操作系统的 MySQL 进行安装,限于篇幅,在此不再介绍 MySQL 安装过程。

在 Spring Boot 2 项目的 pom.xml 文件中引入 MySQL 驱动相关依赖,如下所示:

```xml
<dependency>
    <groupId>mysql</groupId>
    <artifactId>mysql-connector-java</artifactId>
</dependency>
```

下面是配置 MySQL 和 MyBatis 完整的依赖:

```xml
<dependencies>
    <dependency>
        <groupId>org.springframework.boot</groupId>
        <artifactId>spring-boot-starter</artifactId>
    </dependency>
    <dependency>
        <groupId>org.mybatis.spring.boot</groupId>
        <artifactId>mybatis-spring-boot-starter</artifactId>
        <version>2.1.0</version>
    </dependency>
    <dependency>
        <groupId>mysql</groupId>
        <artifactId>mysql-connector-java</artifactId>
    </dependency>
</dependencies>
```

编辑项目配置文件 application.properties,加入下面配置信息:

```
spring.datasource.url=jdbc:mysql://localhost:3306/test?useUnicode=true&characterEncoding=UTF-8&useSSL=false&serverTimezone=GMT%2B8
spring.datasource.username=root
spring.datasource.password=root
spring.datasource.driver-class-name=com.mysql.cj.jdbc.Driver
```

在上面配置中,spring.datasource.url 中配置 MySQL 所在 IP、端口和数据库等,spring.datasource.username 和 spring.datasource.password 配置访问 MySQL 的用户名和密码,spring.datasource.driver-class-name 配置 MySQL 驱动信息。在项目开发过程中,为了快速创建数据表和插入初始数据,希望在项目启动时,自动创建数据表和加载初始数据,则需要在配置文件中加入下面信息:

```
spring.datasource.schema=classpath:db/schema.sql
spring.datasource.data=classpath:db/data.sql
spring.datasource.initialization-mode=always
spring.datasource.continue-on-error=false
```

上面配置中,前两条配置指示初始创建数据表和插入初始数据文件位置,spring.

datasource.initialization-mode 表示初始化方式，其有三个值可选：
① always：始终执行初始化；
② embedded：只初始化内存数据库（默认值），例如 H2 数据库等；
③ never：不执行初始化。

如果不配置 spring.datasource.initialization-mode 项，默认只初始化内存数据库，这也是在前面章节中，没有配置该项但同样能初始化数据的原因。spring.datasource.continue-on-error 项表示遇到错误时，是否继续执行。

在 schema.sql 文件中，书写创建数据表语句时，需要和 MySQL 创建语法相同，创建示例如下所示：

```
create table user (
    user_id bigint AUTO_INCREMENT,
    user_name varchar(25) not null,
    user_address varchar(255),
    user_sex varchar(10),
    primary key (user_id)
);
```

本节介绍的是 MySQL 和 MyBatis，MySQL 和 JPA 的配置与之类似。同时，前面章节中已经介绍如何具体操作数据，在本章不再作具体介绍。本节完整示例详见本节源码。

7.3　连接 MariaDB 数据库

MariaDB 是 MySQL 的一个分支，主要由开源社区在维护，采用 GPL 授权许可。自 MySQL 被甲骨文公司收购后，其有被闭源风险，因此社区建立 MariaDB 分支来避开这个风险。

建立 MariaDB 的目的是完全兼容 MySQL，包括 API 和命令行，使之能轻松成为 MySQL 的代替品。故很多开发人员知道 MySQL，但不清楚 MariaDB。同时，很多开源社区也逐步采用 MariaDB 替代 MySQL 数据库，包括 CentOS 操作系统的源中，默认已更换为 MariaDB 数据库。同时，MariaDB 经过多年社区维护，在很多性能方面已超越 MySQL。本节简要介绍 Spring Boot 2 如何连接操作 MariaDB。

在 MariaDB 官网下载 MariaDB 数据库，官网地址如下：

https://mariadb.org/

选择相应操作系统对应版本下载，如果是 CentOS 7，则可在命令行下输入下面命令在线安装：

```
yum install mariadb-server
```

本节不对 MariaDB 数据库安装过程作详细介绍。

在 Spring Boot 2 项目的 pom.xml 文件中，加入下面依赖：

```
<dependencies>
    <dependency>
```

```xml
            <groupId>org.springframework.boot</groupId>
            <artifactId>spring-boot-starter</artifactId>
        </dependency>
        <dependency>
            <groupId>org.mybatis.spring.boot</groupId>
            <artifactId>mybatis-spring-boot-starter</artifactId>
            <version>2.1.0</version>
        </dependency>
        <dependency>
            <groupId>org.mariadb.jdbc</groupId>
            <artifactId>mariadb-java-client</artifactId>
        </dependency>
    </dependencies>
```

上面代码中，mariadb-java-client 依赖自动加入 MariaDB 相关驱动。接着编辑项目配置文件 application.properties，加入下面内容：

```
spring.datasource.url=jdbc:mysql://localhost:3306/test?useUnicode=true&characterEncoding=UTF-8&useSSL=false&serverTimezone=GMT%2B8
spring.datasource.driver-class-name=org.mariadb.jdbc.Driver
spring.datasource.username=root
spring.datasource.password=root
```

在上面配置项中，唯一和上一节中连接 MySQL 配置不同的是 spring.datasource.driver-class-name 项，更改为 org.mariadb.jdbc.Driver。

MariaDB 是 MySQL 的一个分支，其在操作等方面很多都很相似，包括连接方式、SQL 语句等基本都相同。尽管采用 MySQL 的驱动同时也能连接 MariaDB 数据库，但在此建议，还是采用 MariaDB 专用连接驱动连接 MariaDB 数据库比较好。

7.4 连接 SQL Server 数据库

SQL Server 是 Microsoft 公司推出的关系型数据库管理系统，其具有使用方便、可伸缩性好与相关软件集成程度高等优点。该数据库全面兼容 Windows 平台，具有图形化操作界面、操作简便、容易学习和掌握等特点。一直以来，开发人员都很青睐在 Windows 平台上使用该数据库。但其最新版本可在 Red Hat Enterprise Linux（RHEL）、SUSE Linux Enterprise Server（SLES）和 Ubuntu 上受支持。同时，其也可作为 Docker 映像提供，可在 Linux 上的 Docker 引擎或用于 Windows/Mac 的 Docker 上运行。

作者本人也使用该数据库开发多个数据库应用系统，使用其提供的图形化界面可完成大部分数据库管理操作。在此，不对 SQL Server 的安装和操作做介绍。

在 Spring Boot 2 中连接 SQL Server，首先在项目的 pom.xml 文件中加入下面依赖：

```xml
    <dependencies>
        <dependency>
            <groupId>org.springframework.boot</groupId>
            <artifactId>spring-boot-starter</artifactId>
```

```xml
        </dependency>
        <dependency>
            <groupId>org.mybatis.spring.boot</groupId>
            <artifactId>mybatis-spring-boot-starter</artifactId>
            <version>2.1.0</version>
        </dependency>
        <dependency>
            <groupId>com.microsoft</groupId>
            <artifactId>sqljdbc4</artifactId>
            <version>3.0</version>
        </dependency>
    </dependencies>
```

编辑项目配置文件 application.properties，加入下面内容：

```
#SQL server
spring.datasource.url=jdbc:sqlserver://127.0.0.1:1433;DatabaseName=test
spring.datasource.username=sa
spring.datasource.password=sa
```

在上面配置中，127.0.0.1:1433 是 SQL Server 所在 ip 和端口，test 是数据库名称。username 和 password 是访问用户名和密码。保存配置，完成连接 SQL Server 数据库的配置。

7.5　连接 Oracle 数据库

Oracle 数据库是当前很成熟的关系型数据库之一，在数据库领域一直处于领先地位，其在各行各业都有很好的应用。具有系统可移植性好、使用方便、功能强和适用于各类大、中、小环境等特点，是一种高效率、可靠性好的适应高吞吐量的数据库解决方案。

下面是在 Spring Boot 2 项目中，配置连接 Oracle 方法。

编辑 pom.xml 文件，加入下面依赖：

```xml
<dependencies>
    <dependency>
        <groupId>org.springframework.boot</groupId>
        <artifactId>spring-boot-starter</artifactId>
    </dependency>
    <dependency>
        <groupId>org.mybatis.spring.boot</groupId>
        <artifactId>mybatis-spring-boot-starter</artifactId>
        <version>2.1.0</version>
    </dependency>
    <dependency>
        <groupId>com.oracle</groupId>
        <artifactId>ojdbc6</artifactId>
        <version>11.2.0.3</version>
    </dependency>
</dependencies>
```

在上面代码中，ojdbc6 是连接 Oracle 相关驱动。编辑项目配置文件 application.properties，加入下面内容：

```
#Oracle
spring.datasource.url=jdbc:oracle:thin:@127.0.0.1:1521/orcl
spring.datasource.username=system
spring.datasource.password=orcl
spring.datasource.driver-class-name=oracle.jdbc.driver.OracleDriver
```

在上面配置项中，url 是 Oracle 安装基本信息。username 和 password 是连接 Oracle 的用户名和密码。driver-class-name 是 Oracle 驱动。保存 application.properties，完成 Spring Boot 2 项目连接 Oracle 数据库的配置。

7.6 连接多数据库

在同一个项目中，会遇到需要连接多个数据源的情况，比如业务数据放在一个数据库，日志数据放在另外一个数据库情况，并有可能是同一个类型数据库，也可能是不同类型数据库，比如 MariaDB、SQL Server 等。在 Spring Boot 2 中，可同时支持连接多个数据源，但需要进行一些配置与编码；并取消项目的自动配置，采用手动配置。下面是配置两个数据源，以 JPA 方式连接 MySQL 数据库案例。

编辑 Spring Boot 2 项目的 pom.xml 文件，加入下面依赖：

```xml
<dependencies>
    <dependency>
        <groupId>org.springframework.boot</groupId>
        <artifactId>spring-boot-starter</artifactId>
    </dependency>
    <dependency>
        <groupId>org.springframework.boot</groupId>
        <artifactId>spring-boot-starter-data-jpa</artifactId>
    </dependency>
    <dependency>
        <groupId>org.springframework.boot</groupId>
        <artifactId>spring-boot-configuration-processor</artifactId>
        <optional>true</optional>
    </dependency>
    <dependency>
        <groupId>mysql</groupId>
        <artifactId>mysql-connector-java</artifactId>
    </dependency>
</dependencies>
```

在上面依赖中，加入 JPA、MySQL 等相关依赖，由于多数据源指向同一个数据库 MySQL，只需加入 MySQL 连接驱动即可。编写项目配置文件 application.properties：

```
#数据源1
spring.datasource.primary.driver-class-name=com.mysql.cj.jdbc.Driver
```

```
spring.datasource.primary.url = jdbc:mysql://127.0.0.1:3306/test2?useUnicode=
true&characterEncoding=UTF-8&allowMultiQueries=true&serverTimezone=GMT%2B8
spring.datasource.primary.username = root
spring.datasource.primary.password = root
spring.datasource.primary.initialize = true
spring.datasource.primary.database-platform = org.hibernate.dialect.MySQL5Dialect

#数据源2
spring.datasource.secondary.driver-class-name = com.mysql.cj.jdbc.Driver
spring.datasource.secondary.url = jdbc:mysql://127.0.0.1:3306/test?useUnicode=
true&characterEncoding=UTF-8&allowMultiQueries=true&serverTimezone=GMT%2B8
spring.datasource.secondary.username = root
spring.datasource.secondary.password = root
spring.datasource.secondary.initialize = true
spring.datasource.secondary.database-platform = org.hibernate.dialect.MySQL5Dialect

spring.jpa.hibernate.ddl-auto = update
spring.jpa.show_sql = true
spring.jpa.properties.hibernate.format_sql = true
```

上面配置项中，建立两个数据库连接，自定义前缀分别是 spring.datasource.primary 和 spring.datasource.secondary，这两个前缀根据需要可更换为任意字符串，连接到本地 MySQL 的不同数据库。spring.jpa 开头的配置用于设置是否创建和更新数据表，以及是否打印输出 SQL。创建两个自定义配置类 PrimaryDataSourceConfig 和 SecondaryDataSourceConfig，这两个配置类比较相似，以 PrimaryDataSourceConfig 为例：

```java
@Configuration
@EnableJpaRepositories(
        basePackages = "com.zioer.repositories.primary",
        entityManagerFactoryRef = "primaryEntityManagerFactory",
        transactionManagerRef = "primaryTransactionManager"
)
public class PrimaryDataSourceConfig {
    @Autowired
    private Environment env;

    @Bean
    @ConfigurationProperties(prefix = "spring.datasource.primary")
    public DataSourceProperties primaryDataSourceProperties() {
        return new DataSourceProperties();
    }

    @Bean
    public DataSource primaryDataSource() {
        return primaryDataSourceProperties().initializeDataSourceBuilder().build();
    }

    @Bean
    public PlatformTransactionManager primaryTransactionManager()
    {
```

```java
        EntityManagerFactory factory = primaryEntityManagerFactory().getObject();
        return new JpaTransactionManager(factory);
    }

    @Bean
    public LocalContainerEntityManagerFactoryBean primaryEntityManagerFactory()
    {
        LocalContainerEntityManagerFactoryBean factory = new
        LocalContainerEntityManagerFactoryBean();
        factory.setDataSource(primaryDataSource());
        factory.setPackagesToScan(new String[]{"com.zioer.model.primary"});
        factory.setJpaVendorAdapter(new HibernateJpaVendorAdapter());

        Properties jpaProperties = new Properties();
        jpaProperties.put("hibernate.hbm2ddl.auto",
            env.getProperty("spring.jpa.hibernate.ddl-auto"));
        jpaProperties.put("hibernate.show_sql",
            env.getProperty("spring.jpa.show_sql"));
        jpaProperties.put("hibernate.format_sql",
            env.getProperty("spring.jpa.properties.hibernate.format_sql"));
        jpaProperties.put("hibernate.dialect",
            env.getProperty("spring.datasource.primary.database-platform"));
        factory.setJpaProperties(jpaProperties);

        return factory;
    }

    @Bean
    public DataSourceInitializer primaryDataSourceInitializer()
    {
        DataSourceInitializer dataSourceInitializer = new DataSourceInitializer();
        dataSourceInitializer.setDataSource(primaryDataSource());
        ResourceDatabasePopulator databasePopulator = new
            ResourceDatabasePopulator();
        databasePopulator.addScript(new ClassPathResource("db/primary-data.sql"));
        dataSourceInitializer.setDatabasePopulator(databasePopulator);
        dataSourceInitializer.setEnabled(
            env.getProperty("spring.datasource.secondary.initialize",
            Boolean.class, false));
        return dataSourceInitializer;
    }
}
```

上面代码比较长,其实并不复杂,一是指定扫描资源所在目录及模型所在目录,因为涉及多个数据源,最好将不同数据源的内容放在不同目录下,便于管理。其次,读取 spring.datasource.primary 开头配置项,创建新的数据源等。

创建数据模型方法如下:

User 类对应 spring.datasource.primary 开头数据源,Course 类对应 spring.datasource.secondary 开头数据源。

User.java 文件内容如下：

```java
package com.zioer.model.primary;
@Entity
@Table(name = "USER")
public class User {
    @Id @GeneratedValue(strategy = GenerationType.AUTO)
    private Long id;
    @Column(nullable = false, name = "user_name")
    private String name;
    @Column(nullable = false, name = "user_address")
    private String address;
    @Column(nullable = false, name = "user_sex")
    private String sex;

    //省略 get 和 set 等
}
```

Course.java 文件内容如下：

```java
package com.zioer.model.secondary;
@Entity
@Table(name = "COURSE")
public class Course {
    @Id @GeneratedValue(strategy = GenerationType.AUTO)
    private Long id;
    @Column(nullable = false, name = "course_name")
    private String name;
    @Column(nullable = true, name = "course_desc")
    private String desc;

    //省略 get 和 set 等
}
```

创建资源类文件，UserRepository 接口内容如下：

```java
package com.zioer.repositories.primary;
public interface UserRepository extends JpaRepository<User, Integer>{
}
```

CourseRepository 接口内容如下：

```java
package com.zioer.repositories.secondary;
public interface CourseRepository extends JpaRepository<Course, Integer>{
}
```

下面创建一个简单服务类文件：

```java
package com.zioer.service;

@Service
public class AllService {
```

```java
@Autowired
private UserRepository userRepository;

@Autowired
private CourseRepository courseRepository;

public List<User> getUsers()
{
    return userRepository.findAll();
}

public List<Course> getCourse()
{
    return courseRepository.findAll();
}
}
```

以上代码,完成两个数据源主要代码和调用方法编写。最后需要在主启动文件 Application 中关闭自动配置功能,以启用以上手动配置:

```java
package com.zioer;
@SpringBootApplication(
        exclude = { DataSourceAutoConfiguration.class,
                    HibernateJpaAutoConfiguration.class,
                    DataSourceTransactionManagerAutoConfiguration.class })
@EnableTransactionManagement
public class C71Application implements CommandLineRunner{

    @Autowired
    AllService allService;

    public static void main(String[] args) {
        SpringApplication.run(C71Application.class, args);
    }

    @Override
    public void run(String... args) throws Exception {
        List<User> list = allService.getUsers();
        System.out.println("----------------------");
        System.out.println("-------- users --------");
        for(User user:list) {
            System.out.println(user);
        }

        List<Course> list2 = allService.getCourse();
        System.out.println("----------------------");
        System.out.println("-------- Courses --------");
        for(Course course:list2) {
            System.out.println(course);
        }
    }
}
```

在以上代码中,关键在@SpringBootApplication 注解的配置,其中属性 exclude 排除 DataSourceAutoConfiguration.class、HibernateJpaAutoConfiguration.class 和 DataSource-TransactionManagerAutoConfiguration.class 等。以上项目结构如图 7.1 所示。

图 7.1　示例项目目录

图 7.1 所示项目结构中,将属于不同数据源内容放置在不同目录下即可。尽管该示例项目以连接 MySQL 为例,可根据需要连接不同数据库,比如 H2、SQL Server 等。

本章小结

本章重点介绍了连接几个典型关系数据库的配置方法,最后,以 JPA 为例,介绍了连接不同数据源方法。本章介绍知识点较多,重在实践。

第8章 操作MongoDB

本章将介绍 Spring Boot 2 连接、配置和操作 MongoDB,以及 MongoDB 的基本操作。

8.1 MongoDB 介绍及安装

NoSQL,即不仅仅是 SQL。互联网上各类应用系统产生的大量数据,采用关系型数据库存储和处理,已经不能满足用户需求,比如日志。用户新的想法产生的需求,不能简单、快速用传统方法来描述。NoSQL 概念提得很早,但在互联网高速发展之下,NoSQL 才得以快速发展,很多人提倡用非关系型数据库来存储数据。特别是用户数据成倍增长,我们要对这些数据进行挖掘处理,显然 SQL 已经超出了这个能力,但 NoSQL 能很好弥补和处理。

MongoDB 由 C++编写,是一个基于分布式文件存储的、典型的 NoSQL 数据库。支持的数据结构较松散,即类似 JSON 的 BSON 格式,可以存储比较复杂的数据类型。其语法有点类似于面向对象的查询语言,几乎可以实现类似关系数据库单表查询的绝大部分功能,同时还可以对数据建立索引。

MongoDB 提供 Windows、Linux、Mac 等操作系统进行安装。该数据库设计原则是高性能、可扩展、易部署、易使用,存储数据非常方便。其具有以下特征:

- 模式自由,采用无模式结构存储;
- 支持完全索引,即可以在任意属性上建立索引,包含内部对象;
- 支持查询;
- 提供聚合工具;
- 支持复制和数据恢复功能;
- 可通过网络访问。

本地安装 MongoDB 前,需要在其官网下载,下载地址如下:

https://www.mongodb.com/

下面以在 Windows 操作系统安装为例,在其官网找到下载页面,找到最新版本,目前,其提供最新稳定版本是 4.2,同时,其提供 MSI 和 ZIP 包两种形式。MSI 下载后,直接安装即可使用，建议下载 ZIP 包形式，了解其安装过程。下面以 ZIP 包安装为例，将下载后的 ZIP 包解压到本地任何一个目录，比如 D:\mongodb，如图 8.1 所示。

图 8.1 所示 MongoDB 解压后目录比较简单，实际上就一个 bin 目录。下面进行简单配置，使其运行。在该目录下建立配置文件 mongodb.conf，内容如下：

```
dbpath = D:\mongodb\data
logpath = D:\mongodb\log\mongodb.log
port = 27017
auth = false #是否需要用户名和密码访问
```

图 8.1 MongoDB 解压后目录

以上内容简单配置了数据存放路径、日志所在路径、访问端口（默认 27017 端口，在此列出，允许更改为其他端口）等内容。在该目录下建立批处理文件 run.bat，内容如下：

```
D:\mongodb\bin\mongod.exe -f D:\mongodb\mongodb.conf
```

保存完以后，在 D:\mongodb 目录中，手动建立目录 log 和 data，以存放日志文件和数据文件，这两个目录需要显示存在，MongoDB 不会自动生成。接着双击 run.bat 文件，弹出命令提示符 cmd 窗口，如图 8.2 所示，表示 MongoDB 已正常运行。

图 8.2 MongoDB 运行状态

打开一个新的命令提示符窗口，输入下面命令：

```
d:\mongodb\bin\mongo
```

进入 MongoDB 运行的命令行模式，输入下面命令：

```
show dbs
```

运行结果如图 8.3 所示。

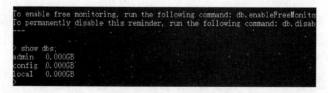

图 8.3 MongoDB 运行命令行模式

以上运行 MongoDB 是采用运行批处理方式，还可以将其注册为系统服务，以设置让 MongoDB 能随计算机启动而启动，简化用户单独启动的烦琐操作。同样，以管理员运行方式打开一个新的终端窗口下，如图 8.4 所示。

在提示符中，输入下面命令：

```
d:\mongodb\bin\mongod.exe -- install -- config "d:\mongodb\mongodb.conf"
```

图 8.4 命令提示符

完成 MongoDB 服务注册，在服务中显示如图 8.5 所示。

图 8.5 MongoDB 服务

在图 8.5 中，可根据需要设置服务是否自动启动，以及启动或停止该服务。如果希望移除该 MongoDB 服务，以管理员身份运行方式打开一个新的终端窗口，输入下面命令即可：

d:\mongodb\bin\mongod.exe - remove

通过以上简单设置，完成了 MongoDB 在 Windows 系统中的安装。

8.2 MongoDB 基本操作

为了更好理解 MongoDB 运行，以及后续在 Spring Boot 2 中操作，本节将介绍 MongoDB 的一些基本知识和操作命令。

在 MongoDB 中，首先需要理解一些概念，这些和关系数据库中的概念对比如表 8.1 所示。

表 8.1 MongoDB 中关键字概念

MongoDB 概念	关系数据库概念	说 明
database	database	数据库
collection	table	集合
document	row	文档
field	column	域
index	index	索引
primary key	primary key	主键

在表 8.1 中，列举了部分 MongoDB 中比较重要概念，比如 collection 表示集合，和关系数据库中 table 表的概念相似，但不再命名为 table；在 MongoDB 中，不再有行的概念，而是叫 document，即文档，这两个概念可以类比，所以，表 8.1 的对比可以方便开发人员理解。

下面命令是和数据库操作相关：

- show dbs：显示当前所有数据库；
- db：显示当前正在使用数据库；
- use local：切换到指定数据库，local 表示数据库名。

在命令提示符下运行效果如图 8.6 所示。

在图 8.6 中，可看到当前数据库名为 test，但在用 show dbs 命令查看当前所有数据库时，是不存在 test 数据库的。实际上，MongoDB 首次运行时，默认当前在 test 数据库，但并不是立即建立，而是在有创建集合或文档时，才会建立该数据库。

图 8.6 运行数据库相关命令

MongoDB中的文档相当于关系数据库中的行,其由一组键值对表示,但不同于关系数据库中行的概念是,其不需要预先定义该文档,即集合没有固定的结构。集合不需要预先定义,当第一个文档插入时,集合就会被创建。在提示符下输入下面命令:

```
use test
db.user.insert({name: '明明',
    sex: '男',
    address: '上海',
    age:12
})
```

以上命令执行后,切换当前数据库为test,第二条命令表示在test数据库中集合user下插入一个文档。如果当前数据库test不存在,则创建,如果集合user不存在,同理,则创建。

以上insert命令,可以使用save命令替换,如下所示,效果一样:

```
db.user.save({name: '萌萌',
    sex: '女',
    address: '重庆',
    age:11
})
```

通过以上操作,在集合user中,插入了两个文档,查看文档命令如下所示:

```
db.user.find()
```

在命令提示符下运行效果如图8.7所示。

```
> db.user.find()
{ "_id" : ObjectId("5d596467abef7ab236ed0c0f"), "name" : "明明", "sex" : "男", "address" : "上海", "age" : 12 }
{ "_id" : ObjectId("5d5965e0abef7ab236ed0c11"), "name" : "萌萌", "sex" : "女", "address" : "重庆", "age" : 11 }
```

图8.7 查看所有文档

图8.7显示了集合user中所有文档。

更新文档命令如下所示:

```
db.user.update({'name':'萌萌'},
    {$set:{'age':12}
})
```

以上update命令表示更新,其中第一个条件类似于SQL中的where条件,第二个参数表示更新,即查找name为萌萌,将age更新为12。

或者用save命令,传入整个文档进行替换,命令如下所示:

```
db.user.save({ "_id" : ObjectId("5d596467abef7ab236ed0c0f"),
    name : "明明",
    sex : "女",
    address : "北京",
    age : 12,
    favcrite : "跑步"
})
```

以上命令整体替换_id为5d596467abef7ab236ed0c0f的文档数据,使用find()命令查看以上两条命令运行结果,如图 8.8 所示。

```
> db.user.find()
{ "_id" : ObjectId("5d596467abef7ab236ed0c0f"), "name" : "明明", "sex" : "女", "address" : "北京", "age" : 12, "favorite" : "跑步" }
{ "_id" : ObjectId("5d5965e0abef7ab236ed0c11"), "name" : "萌萌", "sex" : "女", "address" : "重庆", "age" : 12 }
```

图 8.8　find()查看结果

由图 8.8 可知,两条记录可不对应,第一条多了键值对:
favorite:"跑步"

由此可知,在 MongoDB 中,集合中各文档中键不必一一对应。

对于不再需要的文档,可用下面命令进行删除:

db.user.remove({'name':'萌萌'})

以上采用 remove 命令,其中参数类似于 SQL 中的 where 条件,即删除 name 为萌萌的记录。如果要删除全部文档,采用如下命令:

db.user.remove({})

以上是在 MongoDB 中简单查询全部记录、新增、修改和删除命令。对于查找命令 find(),可以采用类似 SQL 中 where 条件进行精确查找。比如精确查找满足条件的命令如下所示:

db.user.find({key1:value1, key2:value2})

上面命令类似于 SQL 中的 where…and…命令。例如:

db.user.find({name:"明明", sex:"女"})

表示条件中的 OR,采用下面命令:

```
db.user.find(
{
    $ or: [
        {key1: value1}, {key2:value2}
    ]
})
```

以上命令中,使用了关键字 $ or。例如:

db.user.find({ $ or:[{"name":"明明"},{"sex": "女"}]})

对于大于、小于等比较,则需要采用 $ gt、$ lt 等关键字,其对于关键字和使用方法如表 8.2 所示。

表 8.2　比较关键字及用法

操作	格式	操作	格式
大于	{<key>:{ $ gt:<value>}}	小于或等于	{<key>:{ $ lte:<value>}}
小于	{<key>:{ $ lt:<value>}}	不等于	{<key>:{ $ ne:<value>}}
大于或等于	{<key>:{ $ gte:<value>}}		

用法示例如下所示:

```
db.user.find({"age":{$lt:11}})
```

以上命令表示在集合 user 中,查找 age 小于 11 的所有文档。

在 MongoDB 中,查询结果排序方法采用 sort({key:1})方法,其中 key 表示排序对应键,1 表示升序,-1 表示降序。示例代码如下所示:

```
db.user.find().sort({"name":1})
```

当一个集合中文档数量巨大时,查找满足条件的文档将耗费时间,此时,可以在某一键上创建索引,使用方法 ensureIndex({key:1}),其中,key 为待创建索引的键,1 为按升序创建索引,-1 为按降序创建索引。示例代码如下所示:

```
db.user.ensureIndex({"name":1})
```

以上代码为键 name 按升序方式创建索引。

通过本节介绍,可基本理解 MongoDB 操作的方法,可对应 SQL 语句比较,以加深印象。

8.3　Spring Boot 2 连接 MongoDB

在 Spring Boot 2 中连接 MongoDB,只需要在 pom.xml 中加入 MongoDB 相关依赖,如下所示:

```xml
<dependency>
    <groupId>org.springframework.boot</groupId>
    <artifactId>spring-boot-starter-data-mongodb</artifactId>
</dependency>
```

Spring Boot 2 自动处理和加载相关依赖,如果 MongoDB 安装到本机,并采用默认设置,Spring Boot 2 已自动加载默认配置项,该项目已可正常连接并使用 MongoDB 数据库,并默认连接至 test 数据库。当启动空项目时,Console 中提示如下:

```
Monitor thread successfully connected to server with description ServerDescription{address =
localhost:27017, type = STANDALONE, state = CONNECTED, ok = true, version = ServerVersion
{versionList = [4, 2, 0]}, minWireVersion = 0, maxWireVersion = 8, maxDocumentSize = 16777216,
logicalSessionTimeoutMinutes = 30, roundTripTimeNanos = 3066854}
```

如果是作为测试和开发,以上采用 Spring Boot 2 的自动配置是很快捷和方便,但是在实际生产环境,为了安全,一般不建议采用默认配置。所以,在项目配置文件 application.properties 中加入连接相关信息,示例代码如下所示:

```
spring.data.mongodb.uri=mongodb://localhost:27017/test
```

以上连接配置可理解为将 Spring Boot 2 的自动配置展示到配置文件中,其中 localhost 是 MongoDB 数据库所在位置,27017 为 MongoDB 数据库端口号,test 为连接到的数据库

名。显然配置的好处是,可根据实际情况进行修改。当 MongoDB 设置了登录用户名和密码后,配置信息更改为下面代码:

```
spring.data.mongodb.uri=mongodb://username:password@localhost:27017/test
```

在上面配置中,username 为登录用户名,password 为登录密码。通过以上灵活配置,可实现 MongoDB 放置到本地计算机,或安装至专门服务器,并加设登录用户名和密码,提高其安全性。

提示:MongoDB 默认安装情况下,访问是不需要用户名和密码的。

完整的 pom.xml 中依赖如下所示:

```xml
<dependencies>
    <dependency>
        <groupId>org.springframework.boot</groupId>
        <artifactId>spring-boot-starter</artifactId>
    </dependency>
    <dependency>
        <groupId>org.springframework.boot</groupId>
        <artifactId>spring-boot-starter-data-mongodb</artifactId>
    </dependency>
</dependencies>
```

以上空项目及运行效果详见本节源码。

8.4 使用 MongoTemplate 操作

Spring Boot 2 提供了 MongoTemplate 操作 MongoDB 数据库,即采用了面向对象方式,减少了开发人员直接操作 MongoDB 数据库的烦琐。下面介绍 MongoTemplate 的使用方法。

首先,在项目中需要建立业务模型,示例代码如下所示:

```java
package com.zioer.model;

import org.springframework.data.annotation.Id;
import org.springframework.data.mongodb.core.index.Indexed;
import org.springframework.data.mongodb.core.mapping.Document;
import org.springframework.data.mongodb.core.mapping.Field;

@Document(collection = "USER")
@CompoundIndex(name = "idx_name_age", def = "{'userName': 1, 'userAge': -1}")
public class User {
    @Id
    private String id;
    @Field("name")
    @Indexed
    private String userName;
    @Field("phone")
```

```
    private String userPhone;
    @Field("address")
    private String userAddress;
    @Field("sex")
    private String userSex;
    @Field("age")
    private int userAge;

    //省略 get 和 set 等
}
```

在上面建立的业务模型代码中，涉及如下注解：

(1) @Document：将一个 java 类声明为 MongoDB 的文档，可通过 collection 参数重新指定这个类对应的文档。

(2) @Id：设置主键，在定义为主键的列名上标注。如果在模型中不设置 @Id 主键，则 MongoDB 会自动生成一个唯一主键。

(3) @Field：设置一个域，并重新制定域名；但可以省略该注解，则以参数名为域名。

(4) @Indexed：设置该字段需要加索引。其中，可以设置如下属性：

① unique：设置为 true，则表示唯一索引；

② name：设置索引的名称；否则，名称将自动为该字段生成名称；

③ direction：设置索引的排序顺序，默认是升序，值可为 IndexDirection.DESCENDING、IndexDirection.ASCENDING；

④ background：设置为 true 时，索引将在后台应用，并允许在构建索引时进行读写操作；

⑤ useGeneratedName：设置为 true 时，其将忽略 name 属性中给定的索引名称，并使用 MongoDB 生成的名称；

⑥ sparse：如果设置为 true，则索引将跳过缺少索引字段的任何文档；

⑦ expireAfterSeconds：设置集合过期的秒数。默认为 -1 表示没有到期。

示例如下：

```
@Indexed(direction = IndexDirection.ASCENDING, background = true)
private String userName;
```

(5) @CompoundIndex：建立复合索引；一个与多个复合索引的情况如下：

当一个文档只有一个复合索引时，写法如下：

```
@CompoundIndex(name = "idx_name_age", def = "{'userName': 1, 'userAge': -1}")
```

其中，属性 name 表示复合索引名称，def 表示复合索引的域名和排序顺序。

当一个文档要建立多个复合索引时，写法如下：

```
@CompoundIndexes({
    @CompoundIndex(name = "idx_1", def = "{'userName': 1, 'userAge': -1}"),
    @CompoundIndex(name = "idx_2", def = "{'userAddress': 1, 'userSex': -1}")
})
```

如上代码所示，建立多个复合索引时，需要将多个 @CompoundIndex 放在

@CompoundIndexes 注解中。

(6) @Transient：被该注解标注的，将不会保存至 MongoDB 数据库。

(7) @DBRef：关联到另一个 document 对象。

建立完业务模型，下面建立业务处理服务接口：UserService，示例代码如下所示：

```
package com.zioer.service;
public interface UserService {
    void save(User entity);
    void saveAll(List<User> entities);
    void update(User entity);
    void delete(Serializable... ids);
    User find(Serializable id);
    List<User> findAll();
    public List<User> findByConditon(String param, Object value);
    int countByCondition(String[] params,Object[] values);
}
```

以上代码实现在 MongoDB 数据库中的数据增加、修改、删除和查询操作。下面是具体实现代码解析：

```
package com.zioer.service.impl;
@Service
public class UserServiceImpl implements UserService {
    @Autowired
    protected MongoTemplate mongoTemplate;

    public Class<User> getEntityClass(){
        return User.class;
    }

    @Override
    public void save(User entity) {
        mongoTemplate.save(entity);
    }
}
```

上面代码中，定义类 UserServiceImpl，实现接口 UserService，并在该类上标记 @Service 注解，加入自动装配 MongoTemplate 实例，即可使用 MongoTemplate 提供的操作 MongoDB 方法，下面分别进行介绍：

```
mongoTemplate.save(entity)
```

表示保存一个对象到数据库中，save()方法提供有更新和插入两种功能，如果通过传入的关键字 id 找到一个已经存在的文档，则更新文档，如果没有传入关键字 id 参数或者找不到存在相关文档，那么插入一个新文档。

如果仅仅是插入一个文档，可以用 insert 方法，示例代码如下：

```
mongoTemplate.insert(entity);
```

其中，save 和 insert 方法可以带一个 collectionName 参数，指定存入的集合，例如：

```
mongoTemplate.insert(entity,"student");
```

上面代码表示将 entity 存入指定的 student 集合中。

下面方法是同时批量保存多条记录:

```
@Override
public void saveAll(List<User> entities) {
    mongoTemplate.insertAll(entities);
}
```

上面代码中,MongoTemplate 提供了 insertAll 方法,允许同时保存多个文档,以提高效率。

MongoTemplate 还提供了多种更新文档的方法,简单列举如下:

- Upsert:即 update+insert,如果根据条件没有找到对应的数据,则执行插入操作;
- Update:更新操作;
- updateFirst:更新符合条件的第一条记录;
- updateMulti:如果根据查询条件找到对应的多条记录,全部进行更新;
- findAndModify:查询然后进行更新。

下面是代码示例:

```
Query query = new Query();
query.addCriteria(Criteria.where("id").is("5d5beca55af9627b60f69091"));
Update update = new Update();
update.set("userAddress","Shanghai");
mongoTemplate.findAndModify(query,update,User.class);
```

在上面的示例代码中,需要先定义 Query,加入查询规则,然后加入需要更新的内容,上面示例中加入一项更新内容 userAddress,调 findAndModify 方法实现更新。更新时,代码中加入文档中需要更新的各个域和值,更新过程简单明了。一种更加通用的方法能自动匹配更新内容,下面是通用代码示例:

```
@Override
public void update(User entity) {
    Map<String, Object> map = null;
    try {
        map = parseEntity(entity);
    }catch (Exception e) {
        e.printStackTrace();
    }
    String id = null;
    Object value = null;
    Update update = new Update();
    if(map!= null && map.size()>0) {
        for(String key:map.keySet()) {
            if(key.startsWith("{")) {
                id = key.substring(key.indexOf("{")+1,key.indexOf("}"));
                value = map.get(key);
            }else {
```

```
            update.set(key, map.get(key));
        }
    }
}
    mongoTemplate.updateFirst(new Query().addCriteria(Criteria.where(id).is(value)),
update, User.class);
}
```

在上面代码中,首先是获得传入变量 User 中各键值对,采用循环赋值方法进行叠加,最后调用 updateFirst 进行更新操作。下面是查询方法:

```
@Override
public User find(Serializable id) {
    return mongoTemplate.findById(id, User.class);
}
```

上面代码通过关键字 id,返回指定的文档。下面代码是返回全部文档:

```
@Override
public List<User> findAll() {
    return mongoTemplate.findAll(User.class);
}
```

在上面代码中,findAll 方法中无需传递查询条件,即返回全部文档。下面代码示例如何传递查询条件:

```
@Override
public List<User> findByConditon(String param, Object value) {
    Query query = new Query();
    query.addCriteria(Criteria.where(param).is(value));
    return mongoTemplate.find(query, User.class);
}
```

在上面代码中,事先定义 query,通过 Criteria 加入查询规则,由 find 方法返回指定条件的文档集。如果需要删除文档,则通过 remove 等方法进行删除操作,示例代码如下所示:

```
Query query = new Query();
query.addCriteria(Criteria.where("id").is("5d5beca55af9627b60f69091"));
mongoTemplate.findAndRemove(query, User.class);
```

上面代码事先定义 query,加入查询规则,调用 findAndRemove 方法删除文档。

其中,Query 是一个比较重要的查询对象,定义一个查询的所有要素,包括筛选条件、排序、起始行、返回条数等内容。常用方法如表 8.3 所示。

表 8.3 Query 常用方法

方　　法	介　　绍
query(CriteriaDefinition criteriaDefinition)	通过注入 CriteriaDefinition 条件对象获得 Query 查询对象的静态方法:mongoTemplate.find(new Query(Criteria.where("age").lt(13)), User.class);
addCriteria(CriteriaDefinition criteriaDefinition)	添加一个 CriteriaDefinition 查询条件类到 query

续表

方 法	介 绍
skip(long skip)	跳过文档的数量,与 limit 配合使用可实现分页
limit(int limit)	查询返回的文档数量
with(Sort sort)	增加一个 Sort 排序对象
with(Pageable pageable)	增加一个 Pageable 分页对象。同时,Pageable 可以注入一个 Sort,即分页和排序可以一起添加：Pageable pageable=new PageRequest(1,3,sort);

Criteria 属于查询条件类,使用此类定义查询时的查询条件相当于 SQL 的 where。常用方法如表 8.4 所示。

表 8.4 Criteria 常用方法

方 法	介 绍
where(String key)	静态方法,定义查询条件
and(String key)	与操作
gt(Object o)	大于
gte(Object o)	大于或等于
in(Object...o)	包含
is(Object o)	等于
lt(Object o)	小于
lte(Object o)	小于或等于
not()	非
regex(String re)	正则表达式
andOperator(Criteria... criteria)	创建与操作
orOperator(Criteria... criteria)	创建或操作

通过上面代码介绍,MongoTemplate 提供了操作 MongoDB 数据库中数据常用方法。开发人员可以很方便通过这些方法完成业务模型和 MongoDB 数据库中数据间的映射、增加、修改、删除和查询等操作。简化了直接操作 MongoDB 数据库的烦琐,以及可能造成的风险。本节提供了两个源码：一个是上面代码的实现,第二个是进一步封装上面源码,形成更加通用的方法供调用。

8.5 使用 MongoRepository 接口操作

MongoRepository 封装了对 MongoDB 数据库中数据的基本操作,包括以下基本方法：
- findAll()：返回所有文档；
- findAll(Example<S> example)：根据条件,返回文档集,封装为 Example 类型；
- findAll(Example<S> example, Sort sort)：根据条件和排序规则,返回文档集；
- findAll(Sort sort)List<T>：根据排序规则,返回文档集；
- insert(Iterable<S> entities)<S extends T>：插入一个文档；

- insert(S entity):插入一个文档;
- save(S entity):保存一个文档;
- saveAll(Iterable<S> entities):保存所有文档。

下面创建一个资源接口 UserRepository,其继承自 MongoRepository,示例代码如下所示:

```
package com.zioer.repositories;
import org.springframework.data.mongodb.repository.MongoRepository;
public interface UserRepository extends MongoRepository<User, String>{
}
```

以上资源接口具备了常用的 MongoDB CRUD 数据操作,下面建立服务接口 UserService,使其应具备一些常用方法,示例代码如下:

```
package com.zioer.service;

public interface UserService {
    public void insert(User user);                //创建一个新文档
    public void update(User user);                //更新一个文档
    public void delete(User user);                //删除一个文档
    public void deleteAll();                      //删除所有文档
    public User find(User user);                  //根据实例,查找文档
    public User findById(String id);              //根据 id,查找文档
    public List<User> findAll();                  //返回所有文档
}
```

下面创建类 UserServiceImpl,实现上面的接口,示例代码如下所示:

```
package com.zioer.service.impl;

@Service
public class UserServiceImpl implements UserService {
    @Autowired
    UserRepository userRepository;
    @Override
    public void insert(User user) {
        userRepository.insert(user);
    }
    @Override
    public void update(User user) {
        userRepository.save(user);
    }
    @Override
    public void delete(User user) {
        userRepository.delete(user);
    }
    @Override
    public void deleteAll() {
        userRepository.deleteAll();
    }
```

```java
    @Override
    public User find(User user) {
        Example<User> ex = Example.of(user);
        Optional<User> opt = userRepository.findOne(ex);

        return opt.isPresent()?opt.get():null ;
    }
    @Override
    public User findById(String id) {
        Optional<User> opt = userRepository.findById(id);
        return opt.isPresent()?opt.get():null ;
    }
    @Override
    public List<User> findAll() {
        return userRepository.findAll();
    }
}
```

以上代码创建类 UserServiceImpl 并实现接口 UserService，该类用@Service 注解标记为服务。在该类中，自动装配 UserRepository，并在所有方法中，调用 userRepository 中已有方法，完成对于 MongoDB 数据库中数据的基本操作，比如：insert(user)方法实现文档插入，delete(user)方法删除指定文档，findAll()方法返回所有文档等。以上所有方法，开发人员无需再编码实现，只需要调用即可。MongoRepository 在很大程度上简便了开发工作量，提高开发效率。

在查询的高级编程方面，MongoRepository 提供了丰富的接口，以满足开发人员不同的开发需求。比如，在资源接口 UserRepository 中定义下面方法：

```java
public List<User> findByUserName(String username);
```

以上方法将生成查询表达式为{"name"：username }，UserName 在模型文件 User 中对应的域为 name。以上类似接口，可方便开发人员生成各种有用的查询，而无需手动编码实现，表8.5列举了查询方法支持的关键字。

表8.5 支持的查询关键字

关 键 字	简 单 示 例	对应 MongoDB 原生语句
After	findByDateAfter(Date date)	{"date"：{"$gt"：date}}
GreaterThan	findByAgeGreaterThan(int age)	{"age"：{"$gt"：age}}
GreaterThanEqual	findByAgeGreaterThanEqual(int age)	{"age"：{"$gte"：age}}
Before	findBydateBefore(Date date)	{"date"：{"$lt"：date}}
LessThan	findByAgeLessThan(int age)	{"age"：{"$lt"：age}}
LessThanEqual	findByAgeLessThanEqual(int age)	{"age"：{"$lte"：age}}
Between	findByAgeBetween(int from,int to)	{"age"：{"$gt"：from,"$lt"：to}}
In	findByAgeIn(Collection ages)	{"age"：{"$in"：[ages...]}}
NotIn	findByAgeNotIn(Collection ages)	{"age"：{"$nin"：[ages...]}}
IsNotNull,NotNull	findByNameNotNull()	{"Name"：{"$ne"：null}}
IsNull,Null	findByNameNull()	{"Name"：null}

续表

关 键 字	简 单 示 例	对应 MongoDB 原生语句
Like, StartingWith, EndingWith	findByNameLike(String name)	{"Name": name}（name 是正则表达式）
NotLike, IsNotLike	findByNameNotLike(String name)	{"name": {"$not": name}}（name 是正则表达式）
Containing on String	findByNameContaining(String name)	{"name": name}（name 是正则表达式）
NotContaining on String	findByNameNotContaining(String name)	{"name": {"$not": name}}（name 是正则表达式）
Containing on Collection	findByNoteContaining(Address address)	{"note": {"$in": note}}
NotContaining on Collection	findBynoteNotContaining(Address address)	{"note": {"$not": {"$in": note}}}
Regex	findByNameRegex(String firstname)	{"name": {"$regex": name}}
(No keyword)	findByName(String name)	{"name": name}
Not	findByNameNot(String name)	{"name": {"$ne": name}}
Near	findByLocationNear(Point point)	{"location": {"$near": [x,y]}}
Near	findByLocationNear(Point point, Distance max)	{"location": {"$near": [x,y], "$maxDistance": max}}
Near	findByLocationNear(Point point, Distance min, Distance max)	{"location": {"$near": [x,y], "$minDistance": min, "$maxDistance": max}}
Within	findByLocationWithin(Circle circle)	{"location": {"$geoWithin": {"$center": [[x,y], distance]}}}
Within	findByLocationWithin(Box box)	{"location": {"$geoWithin": {"$box": [[x1,y1], x2,y2]}}}
IsTrue, True	findByActiveIsTrue()	{"active": true}
IsFalse, False	findByActiveIsFalse()	{"active": false}
Exists	findByLocationExists(boolean exists)	{"location": {"$exists": exists}}

通过表 8.5，可以灵活编写出需要的查询，例如：

```
public List<User> findByUserAgeGreaterThan(int age);
```

上面代码实现了查询年龄大于指定 age 的所有 User。尽管 MongoRepository 提供了以上丰富的查询功能，如果还不能满足使用情况，或是开发人员不习惯这种方式，喜欢原生查询方式，则 MongoRepository 提供了 @Query 注解，满足这样的需求。下面是简单示例：

```
@Query("{ 'name' : ?0 }")
List<User> findByTheUsersName(String name);

@Query("{name: { $regex: ?0 } })")
List<User> findUserByRegExName(String name);
```

以上代码，通过 @Query 注解，提供了原生 MongoDB 查询方法，方法 findByTheUsersName() 查询域 name 为指定值的所有 User 文档，方法 findUserByRegExName() 查询域 name 中包

含参数值的文档集。通过这样一种方式，能更适合表达复杂的表达式，以及适合于喜欢用原生查询方法的开发人员，同时能更接近于理解MongoDB查询方法。通过@Query注解还能返回指定域值，如下代码所示：

```
@Query(value = "{ 'name' : ?0 }", fields = "{ 'userName' : 1, 'userAge' : 1}")
List < User > findByTheUsersName(String name);
```

以上代码返回指定域userName和userAge，其余的域值将为null。本节的源码结构如图8.9所示。

图8.9　本节工程结构

通过以上的介绍，MongoRepository接口提供了通用的能快速掌握的CRUD方法，能满足于各类开发人员。以上示例详见本节提供的源码。

本章小结

本章知识点较多，涉及MongoDB的介绍、安装和基本使用方法，其次，介绍了Spring Boot 2如何快速配置、连接MongoDB数据库，并详细介绍两种操作MongoDB方法，即MongoTemplate和MongoRepository。在实际开发中，应结合具体案例分析，采取其中更加适合的方法。

第9章 Spring Boot 2 MVC

本章将结合前面章节知识,介绍 Spring Boot 2 中开发 MVC 应用。

9.1 MVC 介绍

MVC 是当前比较流行的开发模式,即 Model+View+Controller。可理解为 Model 负责应用程序核心,比如数据库记录和业务模型;View 负责显示层,用于展示数据;Controller 负责处理接收用户输入,以及写入数据库记录等。理解 MVC 间关系如图 9.1 所示。

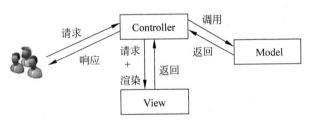

图 9.1 MVC 间关系

MVC 的优点是系统间耦合性低、重用性高、能有效降低生命周期开发成本、实现部署快、后期维护性高,利于软件工程化的管理。特别适合于项目团队的中大型业务系统的研发。

Spring MVC 是基于 MVC 理论进行实现的框架,其实现了松耦合、可插拔等结构,相对于其他 MVC 框架,更能获得广大开发人员的认可,并具有很高的市场占有率。

作者从 Spring 2 开始接触并进行系统研发,至当前 Spring Boot 2,演进非常快,对于开发效率更是一个很大跨越。对于之前为开发人员诟病的 XML 配置,Spring Boot 2 进行了很大简化,加上整合当前流行的 Jar 包管理工具,解决依赖自动加载等,减轻开发人员各种配置,可将注意力更加专注于业务逻辑实现。

9.2　配置 Maven

本章将基于 Spring Boot 2,结合前面各章节知识,建立一个具有 CRUD 操作的 Web 项目,用到的知识点包括:
- MariaDB 数据库;
- Java 1.8;
- Spring Boot 2;
- JPA;
- Thymeleaf。

采用 MariaDB 数据库,相对于前面章节采用 H2,更具有真实性,数据库操作采用 JPA,可以进一步简化代码量;Thymeleaf 是 Spring Boot 2 推荐的前端模板技术。建立项目的第一步是编写 pom.xml 文件,加入以上相关依赖项。pom.xml 文件内容如下所示:

```xml
<?xml version="1.0" encoding="UTF-8"?>
<project
    xmlns="http://maven.apache.org/POM/4.0.0"
    xmlns:xsi="http://www.w3.org/2001/XMLSchema-instance"
    xsi:schemaLocation="http://maven.apache.org/POM/4.0.0
    https://maven.apache.org/xsd/maven-4.0.0.xsd">
    <modelVersion>4.0.0</modelVersion>
    <parent>
        <groupId>org.springframework.boot</groupId>
        <artifactId>spring-boot-starter-parent</artifactId>
        <version>2.2.6.RELEASE</version>
        <relativePath/> <!-- lookup parent from repository -->
    </parent>
    <groupId>com.zioer</groupId>
    <artifactId>c9-1</artifactId>
    <version>0.0.1-SNAPSHOT</version>
    <packaging>jar</packaging>
    <name>c9-1</name>
    <description>demo</description>

    <properties>
        <java.version>1.8</java.version>
    </properties>

    <dependencies>
        <dependency>
            <groupId>org.springframework.boot</groupId>
            <artifactId>spring-boot-starter-thymeleaf</artifactId>
        </dependency>
        <dependency>
            <groupId>org.springframework.boot</groupId>
            <artifactId>spring-boot-starter-web</artifactId>
```

```xml
        </dependency>
        <dependency>
            <groupId>org.springframework.boot</groupId>
            <artifactId>spring-boot-starter-data-jpa</artifactId>
        </dependency>
        <dependency>
            <groupId>org.mariadb.jdbc</groupId>
            <artifactId>mariadb-java-client</artifactId>
        </dependency>
        <dependency>
            <groupId>org.webjars</groupId>
            <artifactId>bootstrap</artifactId>
            <version>4.3.1</version>
        </dependency>
    </dependencies>

    <build>
        <plugins>
            <plugin>
                <groupId>org.springframework.boot</groupId>
                <artifactId>spring-boot-maven-plugin</artifactId>
            </plugin>
        </plugins>
    </build>
</project>
```

上面代码中,关键是依赖部分,加入了 Thymeleaf、JPA、MariaDB 等驱动,以及启动项目需要的相关依赖等,同时加入前端 bootstrap 依赖,方便在页面中直接引用。

9.3 建立模型

下面是建立业务模型的代码,用以描述对象 User。

```java
package com.zioer.model;

@Entity
@Table(name = "user")
public class User {
    @Id
    @GeneratedValue(
            strategy = GenerationType.AUTO,
            generator = "native"
    )
    @GenericGenerator(
            name = "native",
            strategy = "native"
    )
    private Integer id;
```

```
    @NotNull
    @Size(min = 2, max = 30)
    private String name;

    private String phone;
    private String address;
    private String sex;

    @NotNull
    @Min(8)
    private Integer age;

    //省略 get 和 set 等
}
```

在上面代码中,加入了很多注解,包括用于检验的注解,下面是比较常用的检验注解:
- @AssertTrue:注释的元素必须为 true;
- @AssertFalse:注释的元素必须为 false;
- @Min(value):注释的元素必须是一个数字,指定其取值的最小值;
- @Max(value):注释的元素必须是一个数字,指定其取值的最大值;
- @Null:注释的元素必须为 null;
- @NotNull:注释的元素不能为 null;
- @DecimalMin(value):注释的元素必须是一个数字,其值必须大于或等于指定的最小值;
- @DecimalMax(value):注释的元素必须是一个数字,其值必须小于或等于指定的最大值;
- @Size(min=,max=):注释的元素的大小必须在指定的范围内;
- @Digits(integer, fraction):注释的元素必须是一个数字,其值必须在可接受的范围内;
- @Past:注释的元素(日期类型)必须早于当前日期;
- @PastOrPresent:注释的元素(日期类型)必须早于或等于当前日期;
- @Future:注释的元素(日期类型)必须晚于当前日期;
- @FutureOrPresent:注释的元素(日期类型)必须晚于或等于当前日期;
- @Negative:注释的元素必须是负数;
- @NegativeOrZero:注释的元素必须是负数或 0;
- @Positive:注释的元素必须是正数;
- @PositiveOrZero:注释的元素必须是正数或 0;
- @Pattern(regex=,flag=):注释的元素必须符合指定的正则表达式;
- @NotBlank(message =):验证字符串非 null,且长度必须大于 0;
- @Email:注释的元素必须是电子邮箱地址;
- @Length(min=,max=):注释的字符串的大小必须在指定的范围内;
- @NotEmpty:注释的字符串必须非空;

- @Range(min=,max=,message=)：注释的元素必须在合适的范围内。

通过以上检验类注解，可以快速检验数据的有效性，包括用户页面传入数据、系统启动时初始化数据等，有效减少代码量，提高开发效率。

提示：当使用 MySQL、MariaDB 数据库时，采用注解创建 model，为了自动生成主键，可采用下面代码写法：

```java
@Id
@GeneratedValue(
    strategy = GenerationType.AUTO,
    generator = "native"
)
@GenericGenerator(
    name = "native",
    strategy = "native"
)
private Integer id;
```

或者，采用下面方式：

```java
@Id
@GeneratedValue(strategy = GenerationType.IDENTITY)
private Integer id;
```

以上两种方法均可使自动创建的 user 表中 id 为自增类型。

9.4 建立资源及服务

建立资源的目的是，创建自定义的资源继承自 JpaRepository，使其具备数据库表的操作能力，减少代码开发量。代码如下所示：

```java
package com.zioer.repositories;

import org.springframework.data.jpa.repository.JpaRepository;
import com.zioer.model.User;

public interface UserRepository extends JpaRepository<User, Integer>
{
}
```

以上资源接口继承自 JpaRepository，具备数据表的 CRUD 操作能力。由于本示例较简单，没有创建自定义方法。

下面创建服务层，服务层的目的是调用资源层，这是为了独立资源层和服务层，资源层对开发人员不可见，开发人员只需要关注服务层。打个比方，开发人员不关心数据是如何与数据库打交道，可能采用 JPA 方式，也可能采用 MyBatis 方式，以达到解耦。下面是服务层接口代码：

```java
package com.zioer.service;
```

```java
import java.util.List;
import com.zioer.model.User;

public interface UserService {
    public void save(User user);
    public void delete(int id);
    public User findById(int id);
    public List<User> findAll();
}
```

上面接口文件只定义基本接口,以实现数据CRUD。下面代码是实现上面接口的具体方法：

```java
package com.zioer.service.impl;

@Service
public class UserServiceImpl implements UserService {
    @Autowired
    UserRepository userRepository;

    @Override
    public void save(User user) {
        userRepository.save(user);
    }

    @Override
    public void delete(int id) {
        userRepository.deleteById(id);
    }

    @Override
    public User findById(int id) {
        Optional<User> optUser = userRepository.findById(id);
        return optUser.isPresent() ? optUser.get() : null;
    }

    @Override
    public List<User> findAll() {
        Order order1 = new Order(Direction.ASC, "id");
        List<Sort.Order> list = new ArrayList<>();

        list.add(order1);

        Sort sort = Sort.by(list);
        return userRepository.findAll(sort);
    }

}
```

在上面代码中,在类UserServiceImpl上加注解@Service,其中自动装配UserRepository资

源,使得该类中,可以使用 UserRepository 提供的方法。以此,暴露给开发人员只有接口,而隐藏具体实现方法。比如,用户调用 findAll(),采用 JPA 内置方法,返回一个 User 列表。

9.5 建立控制层

控制层和使用用户打交道,接收用户数据,以及调用相应服务,返给用户需要界面。示例代码如下所示:

```java
package com.zioer.controller;

@Controller
public class HomeController implements WebMvcConfigurer
{
    @Autowired UserServiceImpl userService;

    @GetMapping("/")
    public String home(Model model)
    {
        model.addAttribute("users", userService.findAll());
        return "index";
    }

    @GetMapping("/form")
    public String showForm(User user) {
        return "form";
    }

    @PostMapping("/form")
    public String checkUserInfo(@Valid User user, BindingResult bindingResult) {
        if (bindingResult.hasErrors()) {
            return "form";
        }

        userService.save(user);
        return "redirect:/results";
    }

    @GetMapping("/view/{id}")
    public String view(@PathVariable(name = "id") Integer id, Model model) {
        User user = userService.findById(id);
        model.addAttribute("user",user);
        return "view";
    }

    @GetMapping("/edit/{id}")
    public String toEditPage(@PathVariable("id") Integer id, Model model){
        User user = userService.findById(id);
        model.addAttribute("user", user);
```

```
        //到修改页面(form是整合新增/修改的页面);
        return "form";
    }

    @GetMapping("/delete/{id}")
    public String deleteUser(@PathVariable("id") Integer id){
        userService.delete(id);
        return "redirect:/";
    }

    @Override
    public void addViewControllers(ViewControllerRegistry registry) {
        registry.addViewController("/results").setViewName("results");
    }

}
```

上面代码所示示例项目功能比较简单,只需要一个控制类便能完成所有功能。建立类 HomeController,并将其标注为@Controller。在该类中,自动装配 UserServiceImpl,使得该类中所有方法都可调用。该类中用到如下注解:

- @GetMapping:是@RequestMapping(method=RequestMethod.GET)的缩写;
- @PostMapping:是@RequestMapping(method=RequestMethod.POST)的缩写。

同时,该类继承自 WebMvcConfigurer,并覆盖和实现其中的方法 addViewControllers(),实现自定义的路径配置,代码如下:

```
registry.addViewController("/results").setViewName("results");
```

其中 addViewController("/results")是 URL 路径,setViewName("results")是 HTML、模板等页面名称。实际上,以上的写法可用下面方法表达:

```
@GetMapping("/results")
public String results(){
    return "results";
}
```

由于以上方法没有包含任何逻辑,只是实现简单跳转,所以采用 addViewController 更简洁明了。

9.6 建立模板

以上完成业务逻辑处理,下面建立 View 层,即 Thymeleaf 模板,在该项目中,共有 4 个页面。第一个页面 index.html,部分代码如下所示:

```
<head>
<title>首页 - 学生信息</title>
<link rel="stylesheet" th:href="@{/webjars/bootstrap/4.3.1/css/bootstrap.min.css}"/>
```

```html
</head>
<body>
<div class="container">
    <h2 class="text-center" th:text="#{app.title}"></h2>
    <div class="table-responsive" style="text-align:right;margin-bottom:15px">
        <button type='button' onclick="window.location.href='/form'" class="btn btn-primary">新增</button>
    </div>
    <div class="table-responsive">
        <table class="table table-striped table-bordered">
            <thead>
              <tr>
                <th>Id</th>
                <th>姓名</th>
                <th>年龄</th>
                <th>联系地址</th>
                <th>联系电话</th>
                <th>操作</th>
              </tr>
            </thead>
            <tbody>
                <tr th:each="user : ${users}">
                    <td th:text="${user.id}">Id</td>
                    <td th:text="${user.name}">name</td>
                    <td th:text="${user.age}">age</td>
                    <td th:text="${user.address}">address</td>
                    <td th:text="${user.phone}">phone</td>
                    <td>
                        <a class="btn btn-sm btn-success" th:href="@{/view/}+${user.id}">查看</a>
                         
                        <a class="btn btn-sm btn-success" th:href="@{/edit/}+${user.id}">编辑</a>
                         
                        <a class="btn btn-sm btn-danger" th:href="@{/delete/}+${user.id}">删除</a>
                    </td>
                </tr>
            </tbody>
        </table>
    </div>
</div>
</body>
```

以上页面是列表页，主要展示后端传递 users 列表，其中页面样式采用 bootstrap，#{app.title}读取配置文件中定义，`<tr th:each="user：${users}"></tr>`模块循环遍历后端传递 users 列表。展示效果如图 9.2 所示。

第二个页面是新增/编辑页 form.html，实现新增和编辑学生信息，部分代码如下所示：

```html
<form action="#" th:action="@{/form}" th:object="${user}" method="post">
```

图 9.2 列表页

```html
<input type="hidden" name="id" th:value="*{id}">
<table class="table table-striped table-bordered">
 <tr>
  <td>姓名:</td>
  <td><input type="text" th:field="*{name}" /></td>
  <td th:if="${#fields.hasErrors('name')}" th:errors="*{name}">姓名错误</td>
 </tr>
 <tr>
  <td>联系方式:</td>
  <td><input type="text" th:field="*{phone}" /></td>
  <td th:if="${#fields.hasErrors('phone')}" th:errors="*{phone}">联系方式错误</td>
 </tr>
 <tr>
  <td>地址:</td>
  <td><input type="text" th:field="*{address}" /></td>
  <td th:if="${#fields.hasErrors('address')}" th:errors="*{address}">地址错误</td>
 </tr>
 <tr>
  <td>性别:</td>
  <td><input type="text" th:field="*{sex}" /></td>
  <td th:if="${#fields.hasErrors('sex')}" th:errors="*{sex}">性别错误</td>
 </tr>
 <tr>
  <td>年龄:</td>
  <td><input type="text" th:field="*{age}" /></td>
  <td th:if="${#fields.hasErrors('age')}" th:errors="*{age}">年龄错误</td>
 </tr>
 <tr>
  <td colspan="2"><button type="submit" class="btn btn-primary">提交</button>
   
  <button type='button' onclick="window.location.href='/'" class="btn btn-warning">返回</button></td>
 </tr>
```

```
    </table>
  </form>
```

上面代码是 form 表单,用于和后端传递 user 处理相关,其中包含隐藏输入框 id,用来处理新增、编辑状态。在每个 tr 元素中,包含类似下面 td 代码:

th:if="${#fields.hasErrors('phone')}" th:errors="*{phone}"

即判断是否有错误信息,如果有错误提示,将进行显示。运行效果如图 9.3 所示。

图 9.3　新增页面效果

在图 9.3 所示页面,直接单击"提交"按钮,页面显示错误提示,如图 9.4 所示。

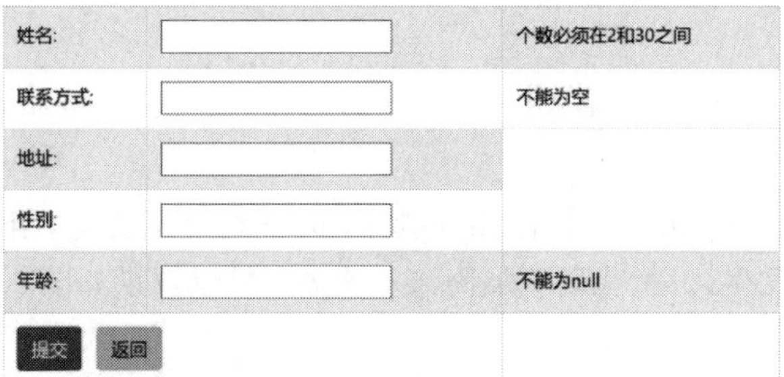

图 9.4　错误提示

由图 9.4 可知,在前面代码中并没有书写代码用于判断等操作,而是由 Spring 的 org.springframework.validation 提供了强有力支持,减少开发人员书写各种判断语句,有效提高开发效率。

第三个页面是展示页面 view.html,部分代码如下所示:

```
<table class="table table-striped table-bordered">
  <tbody>
    <tr>
      <td>ID</td><td th:text="${user!=null}?${user.id}"></td>
    </tr><tr>
      <td>姓名</td><td th:text="${user!=null}?${user.name}"></td>
    </tr><tr>
      <td>年龄</td><td th:text="${user!=null}?${user.age}"></td>
    </tr><tr>
      <td>联系地址</td><td th:text="${user!=null}?${user.address}"></td>
    </tr><tr>
```

```html
        <td>联系电话</td><td th:text = "${user!= null}?${user.phone}"></td>
      </tr>
    </tbody>
</table>
```

该页面比较简单,用 table 表格展示具体学生信息,注意其中写法,需要进行简单判断 user 是否存在,例如:

```html
<td th:text = "${user!= null}?${user.id}"></td>
```

判断 user 是否不为空时,显示其中的 id 值。以上写法,需要在每一个需要展示值的地方都判断,否则会报错;另一种写法,则类似 form.html 页面写法,如下所示:

```html
<table class = "table table-striped table-bordered"
th:if = "${user!= null}" th:object = "${user}">
  <tr>
    <td>ID</td><td th:text = "*{id}"></td>
  </tr><tr>
    <td>姓名</td><td th:text = "*{name}"></td>
  </tr><tr>
    <td>年龄</td><td th:text = "*{age}"></td>
  </tr><tr>
    <td>联系地址</td><td th:text = "*{address}"></td>
  </tr><tr>
    <td>联系电话</td><td th:text = "*{phone}"></td>
  </tr>
</table>
```

在上面代码中,通过 table 元素的 th:if 判断 user 是否不为空,当不为空时,显示 table 中内容,以防止错误发生,以上代码更加简洁。该页面显示效果如图 9.5 所示。

ID	1
姓名	天天
年龄	10
联系地址	北京
联系电话	13122222222

图 9.5 view 个人页面

最后一个页面是静态页,没有任何逻辑处理,只是用于新增或编辑完成后,显示提示。

9.7 系统配置

完成以上代码编写,下面是项目系统配置,打开配置文件 application.properties,增加如下配置项内容:

```
# DataSource Configuration
spring.datasource.driver-class-name=org.mariadb.jdbc.Driver
spring.datasource.url=jdbc:mysql://localhost:3306/test?useUnicode=true&characterEncoding=UTF-8&useSSL=false&serverTimezone=GMT%2B8
spring.datasource.username=root
spring.datasource.password=root

spring.datasource.data=classpath:db/data.sql
# ALWAYS    EMBEDDED    NEVER
spring.datasource.initialization-mode=ALWAYS

# Hibernate Configuration
spring.jpa.hibernate.dialect=org.hibernate.dialect.MariaDB53Dialect
spring.jpa.hibernate.ddl-auto=update
spring.jpa.show-sql=true
# set to false for hot refresh:thymeleaf 缓存(建议：开发环境设置为 false,应用环境设置为 true)
spring.thymeleaf.cache=false
```

在上面配置项中，主要有三部分内容：一是进行数据库配置，包括 MariaDB 驱动、url 项、访问用户名和密码等；其次是设置启动时初始数据，指定初始语句所在文件 db/data.sql，以及必须配置 spring.datasource.initialization-mode，建议第一次运行时设置为 ALWAYS，过后可设置为其他，以防止每次重启时，重复向数据表中增加内容；第三是设置 Hibernate，即设置方言，以及自动创建数据表，设置显示 SQL 语句，利于调试，同时，设置 thymeleaf 不缓存，以使得更改模板后，自动生效。data.sql 中内容为 insert 语句，示例如下所示：

```
INSERT INTO user ( 'name','phone','address','sex','age') VALUES ('天天','13122222222','北京','男',10);
INSERT INTO 'user' ( 'name','phone','address','sex','age') VALUES ('娜娜','13311111111','上海','女',12);
```

初始化内容的优点是利于代码调试和快速看到显示效果。

最后，建立配置文件 messages.properties，内容如下：

app.title=学生信息

该文件的作用是可以被 Thymeleaf 读取，以直接在页面中进行显示。

至此，完成 Spring Boot 2 MVC 项目的开发。

本章小结

本节介绍 Spring Boot 2 建立一个完整的独立 MVC 项目，涉及知识点较多，并加入新的知识点：值校验。值的校验方法很多，包括开发人员写代码校验，借助 Spring 提供的校验方法等。本章的项目结构如图 9.6 所示。

通过本章案例分析，采用 Spring Boot 2 开发 MVC 项目，速度快、效率高，尽管本章介绍是基于 JPA 方式，实际上可以随时替换为 MyBatis、Spring JDBC 等方式，因为结构是解耦的，底层的改变不会对结构上层产生任何影响。建议阅读此章时一定要结合本章提供的源码分析，以加深印象和尽快掌握。

图9.6　本章项目结构

第10章 Spring Boot 2 RESTful

本章介绍 Spring Boot 2 中 RESTful, 以及 Swagger 的应用。

10.1 RESTful 介绍

RESTful 是一种软件架构风格、设计风格, 而不是标准, 只是提供了一组设计原则和约束条件。REST(Representational State Transfer, 表述性状态转移)描述是一个架构样式的网络系统。

当前互联网, 讨论和实现比较多的是 REST 风格架构, 特别是在移动互联网高度发展情况下, 其更容易实现前后端分离, 包括开发、测试、部署等。REST 是一种面向资源架构的应用, 简单理解为所有信息可通过 GET、PUT 等方法进行获取或推送。

第 9 章介绍的 MVC 风格表现出的是结构清晰、前后逻辑和控制解耦, 解决的是服务端的逻辑分层, 前端展示离不开标签库后模板, 尽管可以是松散耦合。相比 MVC 风格, REST 风格注重的是接口统一, 更倾向于跨平台、跨语言实现。REST 解耦的是前端实现和后端逻辑实现, 后端可以采用任何服务端语言, 前端可以采用轻量级语言, 比如纯 HTML 等实现, 以更容易实现跨平台, 在移动端展现。

在 REST 中, 每一个对象都是通过 URL 来表示的, 对象由用户负责将状态信息打包进每一条消息内, 以便对象的处理总是无状态的。设计 RESTful 接口有几个原则:

- 通过 URI(GET, POST, PUT, DELETE)标识资源;
- 通过自我描述的消息;
- 通过表述处理资源;
- 应用状态引擎的超媒体。

当前, Spring 已支持 RESTful API 的开发, 其主要特点是基于 HTTP, 无状态, 主要使用 JSON 作为交互数据(包括入参和响应)。本章将介绍在 Spring Boot 2 下 RESTful 风格的实现。

10.2 Maven 相关配置

本章建立的项目基于如下知识点：
- H2 数据库；
- Java 1.8；
- Spring Boot 2；
- JPA。

由于 RESTful 风格是无页面，故在此无须加入 Thymeleaf 模板，同时本章使用 H2 内存数据库和 JPA 进行演示，建立的 pom.xml 文件中依赖项如下所示：

```xml
<dependencies>
    <dependency>
        <groupId>org.springframework.boot</groupId>
        <artifactId>spring-boot-starter-web</artifactId>
    </dependency>
    <dependency>
        <groupId>org.springframework.boot</groupId>
        <artifactId>spring-boot-starter-data-jpa</artifactId>
    </dependency>
    <dependency>
        <groupId>com.h2database</groupId>
        <artifactId>h2</artifactId>
    </dependency>
</dependencies>
```

保存 pom.xml 文件后，生成的项目中包含的相关 Jar 包很多，如图 10.1 所示。

图 10.1 依赖 Jar 包

通过图 10.1 可知，Spring Boot 2 项目自动加入了 JSON 相关依赖 jackson 等，同时内置 Tomcat、H2 数据库驱动等，省去了开发人员处理 Jar 包相关依赖的工作。

10.3　RESTful API 设计

下面是 model、资源、服务接口和 REST 控制层代码开发。
设计 model，示例代码如下所示：

```java
package com.zioer.model;

@Entity
@Table(name = "user")
public class User {
    @Id
    @GeneratedValue(
            strategy = GenerationType.AUTO,
            generator = "native"
        )
    @GenericGenerator(
            name = "native",
            strategy = "native"
    )
    private int id;
    @NotNull
    private String name;
    private String email;
    private int age;
    //省略 get 和 set
}
```

以上代码为 User 类，作为示例，涉及字段不多，将该 model 设置为 @Entity 注解，以在系统启动时，自动生成数据表。设计资源层接口，代码如下所示：

```java
package com.zioer.repositories;

public interface UserRepository extends JpaRepository<User, Integer>{
}
```

以上代码中，UserRepository 继承自 JpaRepository，使该接口具备基本的 CRUD 功能。下面，设计服务层接口，代码如下所示：

```java
package com.zioer.Service;

public interface UserService {
    public User createUser(User user);
    public List<User> getUser();
    public User findById(int id);
    public User update(User user);
```

```java
    public void deleteUserById(int id);
}
```

以上接口中,设计了几个常用方法,以满足对 User 的 CRUD 操作。创建实现类,示例代码如下所示:

```java
package com.zioer.Service.impl;

@Service
public class UserServiceImp implements UserService {
    @Autowired
    private UserRepository userRepository;

    @Override
    public User createUser(User user) {
        return userRepository.save(user);
    }

    @Override
    public List<User> getUser() {
        return userRepository.findAll();
    }

    @Override
    public User findById(int id) {
        User user = userRepository.findById(id)
                .orElse(null);
        return user;
    }

    @Override
    public User update(User user) {
        return userRepository.save(user);
    }

    @Override
    public void deleteUserById(int id) {
        userRepository.deleteById(id);
    }
}
```

以上类 UserServiceImp 实现接口 UserService 的所有方法,该类中自动装配 UserRepository,并在每个实现的方法中调用 userRepository,完成对 User 的 CRUD 操作。

创建控制层 UserController,以实现 RESTful,代码如下所示:

```java
package com.zioer.rest.controller;

@RestController
@RequestMapping("/api/v1/users")
public class UserController {
    @Autowired
```

```java
    private UserServiceImp userService;

    @GetMapping()
    public List<User> getAllUsers() {
        return userService.getUser();
    }

    @GetMapping("/{id}")
    public  ResponseEntity<User> getUserById(@PathVariable int id)
throws ResourceNotFoundException{
        User user = userService.findById(id);
        if (user == null) {
            new ResourceNotFoundException("User not found on : " + id);
        }

        return ResponseEntity.ok().body(user);
    }

    @DeleteMapping("/{id}")
    public Map<String, String> deleteUser(@PathVariable int id) throws ResourceNotFoundException {
        User user = userService.findById(id);
            if (user == null) {
                new ResourceNotFoundException("User not found on : " + id);
            }

        userService.deleteUserById(id);

        Map<String, String> response = new HashMap<>();
        response.put("status", "deleted");

        return response;
    }

    @PostMapping()
    public ResponseEntity<User> addUser(@Valid @RequestBody User user) {
        User userSaved = userService.createUser(user);

        return new ResponseEntity<User>(userSaved, HttpStatus.CREATED);
    }

    @PutMapping("/{id}")
    public ResponseEntity<User> updateUser(@PathVariable(value = "id") int id,
            @Valid @RequestBody User userDetails)
throws ResourceNotFoundException {
        User user = userService.findById(id);
            if (user == null) {
                new ResourceNotFoundException("User not found on : " + id);
            }

        user.setName(userDetails.getName());
        user.setAge(userDetails.getAge());
```

```
            user.setEmail(userDetails.getEmail());

            User updatedUser = userService.createUser(user);

            return ResponseEntity.ok(updatedUser);
    }
}
```

以上代码中,应用到如下注解:
- @RestController:用于 REST 样式控制器,即处理程序中方法将 JSON/XML 响应直接返回给客户端,而不是使用视图解析器,是 @Controller 和 @ResponseBody 注解的合集,有助于开发人员记忆和调用,及创建 Spring Boot REST 控制器;
- @RequestMapping("/api/v1/users"):该注解用来处理请求地址映射,可作用于类和其中的方法上。当作用于类上,表示该类中的所有响应请求的方法都是以该地址作为父路径,在本例中,即以/api/v1/users 开头。其中,参数 api 表示该 URI 为接口,V1 表示当前版本,users 是具体属于哪个业务,该写法仅是建议;
- @GetMapping():该注解是注解 @RequestMapping(value = "", method = RequestMethod.GET)的简短形式;
- @GetMapping("/{id}"):该注解是注解 @RequestMapping(value = "/{id}", method=RequestMethod.GET)的简短形式;
- @DeleteMapping("/{id}"):该注解是注解 @RequestMapping(value = "/{id}", method=RequestMethod.DELETE)的简短形式;
- @PostMapping():该注解是注解 @RequestMapping(value = "", method = RequestMethod.POST)的简短形式;
- @PutMapping("/{id}"):该注解是注解 @RequestMapping(value = "/{id}", method=RequestMethod.PUT) 的简短形式;
- @PathVariable:该注解用于将路径变量与方法参数绑定;
- @RequestBody:该注解用来接收前端传递给后端的 json 字符串中的数据,同时,前端不能使用 GET 方式进行提交。

通过以上代码编写,实现如下 RESTful API 接口:
- GET 方式:/api/v1/users,用来获得 users 的所有记录;
- GET:/api/v1/ users /{id},通过参数 id,获得某一个用户信息;
- DELETE:/api/v1/ users /{id},通过参数 id,删除用户;
- POST 方式:/api/v1/users,新增一个用户;
- PUT 方式:/api/v1/ users /{id},通过参数 id,更新用户。

创建插入初始数据的文件 data.sql,并将该文件放到目录/src/main/resources/中。其示例如下所示:

```
INSERT INTO 'user'(name,email,age) VALUES ('Tom','tom@163.com', 12);
INSERT INTO 'user'(name,email,age) VALUES ('Kitty','kitty@163.com', 13);
```

该工程结构如图 10.2 所示。

如图 10.2 所示,为了使得返回用户错误信息更加友好,自定义了 exception 显示方式,

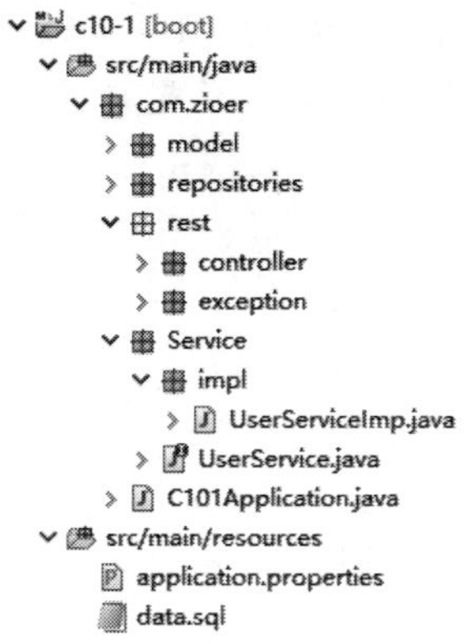

```
v c10-1 [boot]
  v src/main/java
    v com.zioer
      > model
      > repositories
      v rest
        > controller
        > exception
      v Service
        v impl
          > UserServiceImp.java
        UserService.java
      C101Application.java
  v src/main/resources
    application.properties
    data.sql
```

图 10.2　RESTful 示例结构

并放在文件夹 exception 下。

运行以上工程,并在浏览器中输入下面的地址:

$$http://127.0.0.1:8080/api/v1/users/$$

得到如图 10.3 所示结果。

```
[{"id":1,"name":"Tom","email":"tom@163.com","age":12},
{"id":2,"name":"Kitty","email":"kitty@163.com","age":13},
{"id":3,"name":"Jim","email":"jim@163.com","age":11},
{"id":4,"name":"Yammi","email":"yummi@163.com","age":12},
{"id":5,"name":"Rainy","email":"rainy@163.com","age":10},
{"id":6,"name":"Lucy","email":"lucy@163.com","age":13}]
```

图 10.3　浏览器访问 API 接口

图 10.3 展示为全部记录数据,JSON 格式。

至此,完成基本 RESTful API 接口设计,并成功运行。不复杂,但是需要事先规划好各接口,及具备什么样的功能。具体代码参考本节提供的完整源码。

10.4　Swagger 应用

前面介绍了 RESTful 是开发软件架构风格,并没有形成标准和规范,开发人员根据自己所想,可做到随想随写。如果是个人独立开发一个系统,没有任何问题。问题在于,如果是一个团队开发,还是希望有个约束或文档之类的东西,供团队成员理解和学习。

前后端开发,重要连接纽带是 API 接口和相关文档。那么 Swagger 正是我们需要的工具。

Swagger 是一套围绕 OpenAPI 规范构建的开源工具,可以帮助开发人员设计、构建、记录和使用 RESTful API。通过 Swagger 正确定义后,前端开发人员可以使用它快速理解远程服务并与之交互。主要的 Swagger 工具包括:

- Swagger Editor:基于浏览器的在线编辑器,可以在其中编写 OpenAPI 定义;
- Swagger UI:将 OpenAPI 规范呈现为交互式 API 文档;

- Swagger Codegen：从 OpenAPI 规范生成服务器存根和客户端库。

要在 Spring Boot 2 项目中使用 Swagger，首先是在 Maven 配置文件 pom.xml 中增加 Swagger 相关依赖，以 10.3 节中示例为例，在 pom.xml 中，加入如下所示依赖：

```xml
<dependency>
    <groupId>com.spring4all</groupId>
    <artifactId>swagger-spring-boot-starter</artifactId>
    <version>1.9.0.RELEASE</version>
</dependency>
```

这种方法最简单，会直接在项目中加入 Swagger 相关依赖，如图 10.4 所示。

图 10.4　Swagger 相关依赖

接着，需要建立 Swagger 的配置文件，设置一些相关信息等。示例代码如下所示：

```java
package com.zioer;

@Configuration
@EnableSwagger2
public class Swagger2Config {
    /**
     * Swagger2 的配置文件：内容、扫描包等
     *
     * @return the docket
     */
    @Bean
    public Docket createApi() {
        return new Docket(DocumentationType.SWAGGER_2)
                .groupName("public-zioer")
                .apiInfo(apiInfo())
                .select()
                .paths(apiPaths())
//                .paths(PathSelectors.any())
                .build();

    }

    private Predicate<String> apiPaths() {
        return or(regex("/api/v1.*"), regex("/api/v2.*"));
    }
```

```
/**
 * 构建 API 文档的基本信息
 * @return ApiInfo
 */
private ApiInfo apiInfo() {
    return new ApiInfoBuilder()
            .title("学生信息系统 API")    //页面标题
            .description("在该系统中相关 API 接口")   //描述
            .termsOfServiceUrl("http://www.zioer.com")
            .contact("hero803@163.com")    //联系人
            .license("Zioer License")
            .licenseUrl("zioer803@163.com")
            .version("1.0")    //版本号
            .build();
}
```

如以上代码所示,建立类文件 Swagger2Config,在类名上加注解@Configuration 和 @EnableSwagger2,即让 Spring Boot 2 加载该类配置以及启用 Swagger2。

定义 Docket bean 并使用 select()方法获取 ApiSelectorBuilder 的实例,ApiSelectorBuilder 用于配置暴露给 Swagger 公开的接口;apiInfo()用来创建该 API 的基本信息,在其中使用了路径选择谓词。

通过上面的简单配置,重启服务后,在浏览器中输入下面的地址,访问 Swagger UI:

http://localhost:8080/swagger-ui.html#/user-controller

得到如图 10.5 所示的界面。

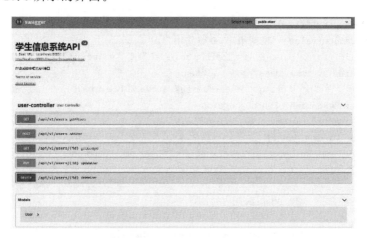

图 10.5　Swagger2 UI 运行界面

在图 10.5 中,能得到项目中所有自定义 API 接口。但这样的文档主要针对请求本身,其描述主要来自于函数等命名产生,对开发人员友好性较差,此时,最好能增加一些说明来丰富文档内容。

需要用到如下注解给相应 API 增加说明:
- @Api:用来标记一个 Controller 类为 Swagger 资源;
- @ApiOperation:用来给 API 增加说明;

- @ApiImplicitParams：用来给参数增加说明集；
- @ApiImplicitParam：用来给参数增加说明。

下面的代码是在控制类 UserController 中应用以上注解，进行说明：

```java
package com.zioer.rest.controller;

@RestController
@RequestMapping("/api/v1/users")
@Api(tags = "用户相关操作", description = "用户操作 API")
public class UserController {

    @GetMapping("/{id}")
    @ApiOperation(value = "根据用户 ID 获取用户信息", notes = "test: 系统初始时,1-6 有正确返回")
    @ApiImplicitParam(paramType = "path", name = "id", value = "用户 ID", required = true, dataType = "Integer")
    public ResponseEntity<User> getUserById(@PathVariable int id)
        throws ResourceNotFoundException{
            User user = userService.findById(id);
            if (user == null) {
                new ResourceNotFoundException("User not found on : " + id);
            }

            return ResponseEntity.ok().body(user);

    }

    @DeleteMapping("/{id}")
    @ApiOperation(value = "根据用户 ID 删除用户")
    @ApiImplicitParam(paramType = "path", name = "id", value = "用户 ID",
        required = true, dataType = "Integer")
    public Map<String, String> deleteUser(@PathVariable int id)
        throws ResourceNotFoundException {
        User user = userService.findById(id);
            if (user == null) {
                new ResourceNotFoundException("User not found on : " + id);
            }

        userService.deleteUserById(id);

        Map<String, String> response = new HashMap<>();
        response.put("status", "deleted");

        return response;
    }

    @PostMapping()
    @ApiOperation(value = "新建用户")
    @ApiImplicitParam(paramType = "RequestBody", name = "user", value = "用户信息",
required = true, dataType = "User")
```

```java
    public ResponseEntity<User> addUser(@Valid @RequestBody User user) {
        User userSaved = userService.createUser(user);

        return new ResponseEntity<User>(userSaved, HttpStatus.CREATED);
    }

    @PutMapping("/{id}")
    @ApiOperation(value = "根据用户ID,更新用户信息")
    @ApiImplicitParams({
        @ApiImplicitParam(paramType = "path", name = "id", value = "用户ID", required = true, dataType = "Integer"),
        @ApiImplicitParam(paramType = "RequestBody", name = "userDetails", value = "用户修改后信息", required = true, dataType = "User")
    })
    public ResponseEntity<User> updateUser(@PathVariable(value = "id") int id,
            @Valid @RequestBody User userDetails) throws ResourceNotFoundException {
        User user = userService.findById(id);
        if (user == null) {
            new ResourceNotFoundException("User not found on : " + id);
        }

        user.setName(userDetails.getName());
        user.setAge(userDetails.getAge());
        user.setEmail(userDetails.getEmail());

        User updatedUser = userService.createUser(user);

        return ResponseEntity.ok(updatedUser);
    }
}
```

以上代码中,在类名、各个方法上都加入了Swagger相关注解,以提高Swagger生成文档的可读性。同时,Swagger提供了对Model层的注解,如下所示:

- @ApiModel:对实体的描述;
- @ApiModelProperty:对实体字段的描述。

在Model层的类User中使用方法如下所示:

```java
package com.zioer.model;

@Entity
@Table(name = "user")
@ApiModel(description = "用户Model", value = "User")
public class User {
    @Id
    @GeneratedValue(
            strategy = GenerationType.AUTO,
            generator = "native"
    )
    @GenericGenerator(
            name = "native",
```

```
            strategy = "native"
)
@ApiModelProperty(value = "用户编号",required = true,dataType = "int")
private int id;
@NotNull
@ApiModelProperty(value = "用户姓名",required = true,dataType = "String")
private String name;
@ApiModelProperty(value = "用户 E-mail",required = false,dataType = "String")
private String email;
@ApiModelProperty(value = "用户年龄",required = false,dataType = "int")
private int age;

}
```

在上面部分代码中,给类名和类的每个字段都加上 Swagger 相关注解,以提高 Swagger 生成文档可读性。编写完以上代码,重新运行工程,再次访问:

$$\text{http://localhost:8080/swagger-ui.html\#/user-controller}$$

页面如图 10.6 所示。

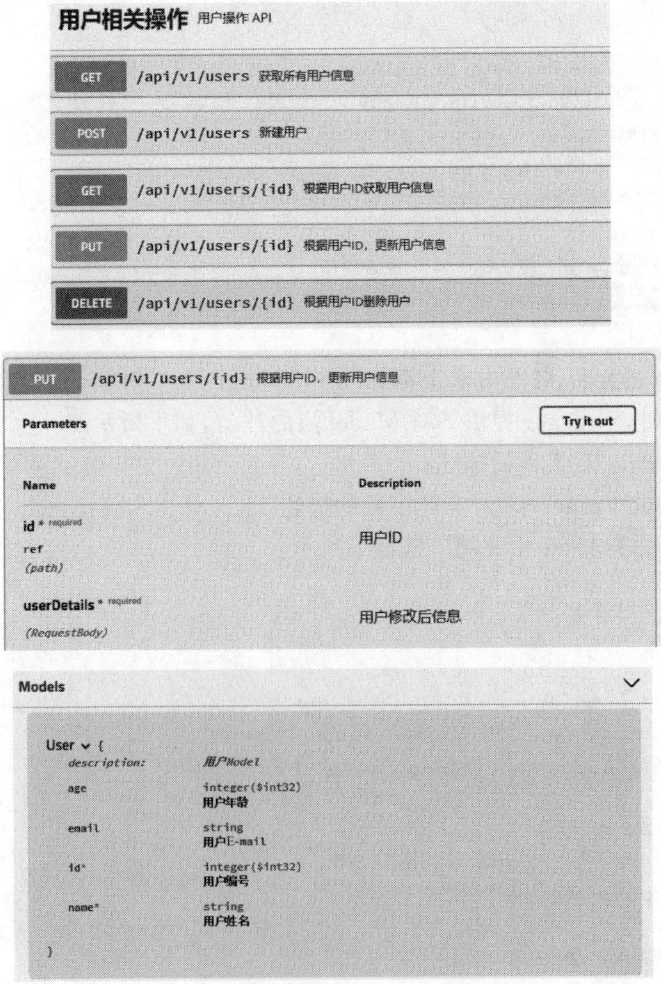

图 10.6 Swagger 运行 UI 示例结果

在图 10.6 中有三个结果：一是用户操作所有 API 的展示及相关描述；二是 PUT 方法中的参数描述；三是 Model 层描述。

通过以上简单示例，可知 Swagger 在对 RESTful API 接口描述方面，提供了定义方式和方法，以及最后自动生成相关 Swagger 文档，供开发人员参考，能有效提高项目的开发进度。本节介绍的方法提供完整源码。

10.5　RESTful API 测试工具

10.4 节进行了 RESTful API 接口设计，但不是每个接口都能通过浏览器地址栏进行访问，比如 POST、PUT 等方式，不能通过浏览器地址栏方式直接访问。此时，需要一款可以及时访问并能测试接口的工具。

目前，用于 RESTful API 接口测试的工具比较多，比如著名的 Chrome 插件、Advance REST Client 插件、RESTful API Client Chrome 插件、DEV HTTP CLIENT 和 Simple REST Client 插件等，这些插件功能各有优点，在 Chrome 中安装后，能较好完成 RESTful API 接口的测试。

本节，推荐另一款免费工具：Postman。该工具也提供 Chrome 插件版，但目前没有再更新。但其提供了本地安装版，并有 Windows、macOS 和 Linux（x64）版，满足各类开发人员适应多种操作环境需求。其官网下载地址如下所示：

　　　　　　　https://www.getpostman.com/downloads/

下面以 Windows 版本为例，下载最新版本到本地后，双击便自动完成安装。其启动界面如图 10.7 所示。

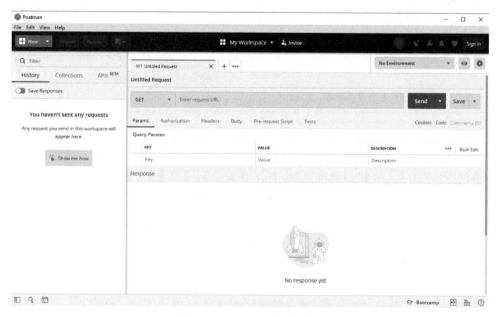

图 10.7　Postman 启动界面

Postman 提供的接口测试功能比较全，包括 GET、POST、PUT 和 DELETE 等方式，同时提供各种参数，文件上传等测试。

只需在 URL 地址栏中输入需要测试的 RESTful API 接口地址，选择 Request 方式，以及输入可能有的参数，然后单击 Send 按钮，即可完成一次 API 调用。比如，测试 POST 方式，在 URL 地址栏中输入"http://127.0.0.1:8080/api/v1/users"。然后，选择 Request 方式为 POST，单击 Body 标签，在其中选择 raw 单选，在数据格式下拉列表中，选择 JSON 格式，在输入框中输入下面内容：

```
{
    "name": "Simeny",
    "email": "simeny@163.com",
    "age": 11
}
```

单击 Send 按钮，运行完成后，结果如图 10.8 所示。

图 10.8 POST 运行结果

作者在开发项目中使用过多个 Chrome 插件、Firefox 插件，以及 Postman 等工具，发现 Postman 不断更新和完善，能适用于开发中的多种场景。最后，希望开发人员能找到一款得心应手的 RESTful API 接口测试工具，对于开发中测试很重要，特别是推送给前端开发人员，能及时了解接口返回结果，便于前端页面设计和开发。

10.6 整合前端

RESTful API 接口开发完成，并通过测试相关工具测试完成。接着，便可整合前端页面，呈现给用户。整合前端有几种方式，其中一种就是直接将前端页面放入同一个工程目录中，比如第 9 章介绍的 MVC 模式，当然，这种方式不是最优的方式。还可采用 RESTful

API 方式，重要的是支持移动应用，移动端可以通过接口调用，并返回数据，包括 PC 端也可轻量化。也可采用纯前端模式开发前端应用，现在涌现出很多新技术支持这种方法，比如前端框架 Vue、Angular 等。这些新技术的出现，可以解决开发中出现的各种前后端问题，实现轻前端、轻后端等开发形式。

如果采用第一种方式，即将前端开发页面放入同一个工程中，就比较好实现，因为在同一个域下，不存在跨域问题。用户通过浏览器访问服务器 A，所有资源都在服务器 A 上，理解为同源：指同一个请求协议（比如：Http 或 Https）、同一个 Ip、同一个端口，3 个全部相同，即为同源。此时，前、后端不需作任何适配，便可访问所有数据。

当前端和后端不再同源时，则前端访问后端数据，包括静态资源时，就属于跨域。此时，前端通过 AJAX 访问后端时，页面报错，如图 10.9 所示。

```
Access to XMLHttpRequest at 'http://localhost:8080/api/v1/users' from origin 'http://127.0.0.1' has (index):1
been blocked by CORS policy: No 'Access-Control-Allow-Origin' header is present on the requested resource.
```

图 10.9　跨域访问错误

那么，就需要进行适配，使得前端通过 AJAX 顺利访问服务端提供资源。现在，跨域的需求很重要，W3C 组织制定了一个 Cross-Origin Resource Sharing 规范，简写为 CORS，其已被大多数浏览器支持。Spring Boot 2 对 CORS 提供了支持，开发人员可以很容易开发出跨域应用。下面介绍在 Spring Boot 2 中如何配置，使得应用支持 CORS。

Spring Boot 2 提供了以下两种方式来支持跨域请求：
- @CrossOrigin 注解：可以定义在某一个控制层的类或方法上，即局部范围应用；
- WebMvcConfigurer 对象：提供全局配置方式。

使用@CrossOrigin 注解方法的示例代码如下所示：

```
@CrossOrigin
public class UserController {
    @Autowired
    private UserServiceImp userService;
    //省略该类中其他内容
}
```

在上面示例中，直接在类 UserController 的类名上加@CrossOrigin 注解，则该类中提供的 RESTful API 接口都能被跨域访问。该注解提供了如下可使用的参数：
- value String 数组：配置允许访问的源；
- origins String 数组：配置允许访问的源；
- allowedHeaders String 数组：配置允许的自定义请求头；
- exposedHeaders String 数组：配置响应的头信息；
- methodsString 数组：配置跨域请求支持的方式，如 GET、POST；
- allowCredentials String：配置是否允许发送 Cookie；
- maxAge long：配置预检请求的有效时间，单位是秒。

下面示例演示如何使用参数：

```
@CrossOrigin(origins = "*",maxAge = 3000, allowCredentials = "false")
public class UserController
```

全局 CORS 配置方法是作用于整个系统，首先建立配置文件类 CorsConfig，示例代码如下所示：

```java
package com.zioer;

@Configuration
public class CorsConfig {
    @Bean
    public WebMvcConfigurer corsConfigurer() {
        return new WebMvcConfigurer() {
            @Override
            public void addCorsMappings(CorsRegistry registry) {
                registry.addMapping("/**");
            }
        };
    }
}
```

在上面代码中，用到了 CorsRegistry 的 addMapping 方法，其是用于配置支持跨域的路径。以上配置方法中，默认情况下所有的域名和 GET、HEAD 和 POST 方法都是允许的。更复杂的配置如下代码所示：

```java
@Override
public void addCorsMappings(CorsRegistry registry) {
    registry.addMapping("/api/**")
        .allowedOrigins("http://127.0.0.1","file://")
        .allowedMethods("GET","PUT", "DELETE")
        .allowedHeaders("header1", "header2", "header3")
        .exposedHeaders("header1", "header2")
        .allowCredentials(false).maxAge(3600);
}
```

以上代码中，用到了下面方法：

- addMappings：配置匹配路径，其中 * 只能匹配到下一层路径，** 表示多层路径；
- allowedOrigins：配置允许的源；
- allowedMethods：配置支持跨域请求的方法，如 GET、POST、PUT、PATCH、DELETE、OPTIONS、HEAD、TRACE 等；
- allowedHeaders：配置用于预检请求的允许的自定义请求头；
- exposedHeaders：配置响应的头信息；
- allowCredentials：配置是否允许发送 Cookie；
- maxAge：配置预检请求的有效时间。

以上两种方法均可解决在 Spring Boot 2 下的跨域请求问题。具体实现请参考 10.5 节中提供的源码。

下面介绍在前端页面中如何使用 AJAX 请求和获取资源。

在 HTML 页面中加入下面代码：

```
< script type = "text/javascript"
```

```
        src = "https://cdn.bootcss.com/jquery/3.4.1/jquery.js"></script>

<script>
    //方法一
    function ajax_get() {
        $.ajax({
            url: "http://localhost:8080/api/v1/users",      //JSON 文件位置
            type: "GET",                                     //请求方式为 GET
            dataType: "JSON",                                //返回数据格式为 JSON
            success: function(data) {                        //请求成功完成后要执行的方法
                //each 循环 使用 $.each 方法遍历返回的数据 date
                $.each(data, function(i, item) {
                    var str = '<div>姓名:' + item.name + ' email: ' + item.email;
                    str += ' 年龄: ' + item.age;
                    str += '</div>';
                    document.write(str);
                })
            }
        })
    }
    ajax_get();                                              //执行
</script>
```

在上面 JavaScript 语句中，AJAX 方法采用 GET 方法获取 JSON 数据，并遍历显示到页面上，比较简单，没做任何样式，直接在页面中显示数据，将上面代码保存至 HTML 页面，运行后，在浏览器中显示如图 10.10 所示。

图 10.10　AJAX 测试

以上 HTML 代码详见本节提供源码。运行方式是，首先运行上一节源码，然后双击本节 HTML 文件，即可看到效果。

本章小结

本章介绍了 Spring Boot 2 下 RESTful API 接口方法，结合案例分析，逐步深入，重点是设计方法，接口统一，同时，介绍了图形化界面测试接口，方便在开发中随时看到效果。还介绍了一个 RESTful API 简单管理插件 Swagger，可以及时查看和管理系统中的接口，其中一个重要作用是，适用于团队开发的前后端对接。在本章最后，介绍了前端访问，以及跨域设置问题，该问题在目前移动互联网开发中显得尤为重要。本章提供源码，请结合源码运行，加深理解本章内容。

第11章
Spring Boot 2安全

本章由浅入深详细介绍 Spring Boot 2 的安全访问技术 Spring Security,便于开发人员结合实际需要进行恰当选择。

11.1 安全介绍

安全是互联网应用的重要一部分组成。只要是应用,就涉及安全性访问、控制等。简单理解就是,一个互联网应用中有增、查、删等基本功能,可能希望匿名用户只有查看权限,只有特定用户有新增的权限,等等。这就需要权限控制,如果开放应用中的所有权限给任何一个人,则可能对系统、甚至计算机造成破坏,这是不希望发生的。

那么,一个互联网应用的安全体系应包含用户认证和用户授权两部分。用户认证即通过用户名和密码,让系统进行认证是否为合法用户,只有合法用户才能访问应用。同理,在一个系统中,不同用户应有不同权限,比如部分用户只有访问列表权限,部分用户有编辑的权限。这就需要对用户进行授权操作,但一般按角色进行授权,然后对用户进行角色分配,一个用户拥有一个或多个角色。

值得关注的是,Spring Security 框架提供了很好的安全支持。开发人员可以很容易地将权限控制放到 Spring Boot 2 的应用中,使 Spring Boot 2 的应用具有更广泛性和可实用性。Spring Security 主要核心功能是认证和授权。在用户认证方面,Spring Security 框架支持主流的认证方式,包括 HTTP 基本认证等;在用户授权方面,Spring Security 提供了基于角色的访问控制和访问控制列表(Access Control List,ACL),能对应用中的领域对象进行细粒度的控制。

本章将结合案例分析,由浅入深介绍在 Spring Boot 2 应用中整合 Spring Security,并提供安全认证的解决方案。

11.2 Spring Boot 2 中快速整合 Spring Security

在 Spring Boot 2 中，整合 Spring Security 框架比较容易。建立一个 Spring Boot 2 工程，编辑 pom.xml 文件，加入下面依赖：

```
<dependencies>
    <dependency>
        <groupId>org.springframework.boot</groupId>
        <artifactId>spring-boot-starter-security</artifactId>
    </dependency>
    <dependency>
        <groupId>org.springframework.boot</groupId>
        <artifactId>spring-boot-starter-web</artifactId>
    </dependency>
</dependencies>
```

在上面代码中，重要的是 spring-boot-starter-security 依赖，该依赖加入 Spring Security 相关 jar 包。至此，实际上已完成了 Spring Security 相关依赖加入，如果不做任何配置，默认情况下，已为应用程序启用身份验证。保存以上 pom.xml 文件，启动应用。在 Console 窗口中，出现如图 11.1 所示内容。

```
2019-09-02 22:19:41.409  INFO 39764 --- [           main] o.s.s.concurrent.ThreadPoolTaskExecutor
2019-09-02 22:19:41.601  INFO 39764 --- [           main] .s.s.UserDetailsServiceAutoConfiguration

Using generated security password: fd4f9985-4fed-4c24-8bc7-36e9eedd7a54

2019-09-02 22:19:41.685  INFO 39764 --- [           main] o.s.s.web.DefaultSecurityFilterChain
```

图 11.1 启动后 Console 窗口内容

其中，Using generated security password 表示自动生成的密码串，用户名默认为 user。在浏览器中输入下面网址进行访问：

http://127.0.0.1:8080

出现如图 11.2 所示界面，自动导向到 login 页面。

图 11.2 login 页面

在图中输入用户名 user，密码为前面自动生成的密码串后，才能进入。当然，进入后是个错误页面，毕竟没有写任何页面代码。

如果想更换默认用户名和随机密码，可以在配置文件application.properties中增加下面配置：

```
spring.security.user.name = test
spring.security.user.password = test
```

至此，已更改测试用的用户名和密码。但以上登录用户名和密码只能用于测试和体验，无法用于实际应用系统中。

11.3 更改自动配置方式

Spring Security自动配置方式默认只有一个用户进行体验，显然不适合一个系统开发。那么，首要的是更改自动配置方式为手动配置，提高其灵活性。下面是更改方法。

打开项目启动类文件，更改启动项类名上的注解@SpringBootApplication，加入属性exclude，如下代码所示：

```java
@SpringBootApplication(exclude = { SecurityAutoConfiguration.class })
public class C111Application {
    public static void main(String[] args) {
        SpringApplication.run(C111Application.class, args);
    }
}
```

以上代码中，在启动配置项中，排除掉Spring Security的自动配置。创建Spring Security的配置类，示例代码如下所示：

```java
package com.zioer;

@Configuration
@EnableWebSecurity
public class SecurityConfiguration extends WebSecurityConfigurerAdapter {

    @Override
    protected void configure(AuthenticationManagerBuilder auth)
        throws Exception {
        auth
            .inMemoryAuthentication()
            .passwordEncoder(new BCryptPasswordEncoder())
            .withUser("user")
            .password(new BCryptPasswordEncoder().encode("123456"))
            .roles("USER")
            .and()
            .withUser("admin")
            .password(new BCryptPasswordEncoder().encode("admin"))
            .roles("USER", "ADMIN");
    }
}
```

```
    @Override
    protected void configure(HttpSecurity http) throws Exception {
        http
            .authorizeRequests()
            .anyRequest()
            .authenticated()
            .and()
            .httpBasic();
    }
}
```

在以上配置类中,应用两个注解@Configuration 和@EnableWebSecurity,标识该类为 Security 配置类。该类只有两个配置方法:第一个配置方法加入了多个用户,并且用户的密码采用加密存储,使用的是内置 BCryptPasswordEncoder()加密方法,加入了两个用户,并指定各用户的角色;第二个配置方法中,为了示例简便化,所有页面都需要访问控制。

保存以上代码,启动服务。在浏览器中打开下面地址进行访问:

$$\text{http://127.0.0.1:8080}$$

出现如图 11.3 所示界面。

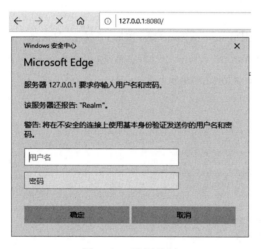

图 11.3 访问控制

在图 11.3 所示弹出窗口中,输入用户名和密码,单击"确定"按钮。提示:输入前面代码中定义的两个用户名和密码中的任意一个均可登录并访问。

通过以上代码设置:一是通过自定义配置文件,可以设置任意多个用户名、密码和角色,并可控制角色访问的细粒度。其次,没有默认的 Web 登录页面,只有一个弹框输入用户名和密码。这是由于取消了 Security 的自动配置的同时,也取消了登录用的默认页面。

11.4 自定义加密配置方式

在 11.3 节中介绍的自定义配置中,加入了下面的加密配置:

```
new BCryptPasswordEncoder()
```

该方法采用的是 SHA-256+随机盐+密钥对密码进行加密,但是有个缺点,直接将其写入到代码中时,如果更换加密方式,则需要同时对多个地方进行修改。那么,一种比较好的习惯是,将加密方式单独进行配置,每次更改加密方式时,则只需要修改一处。下面,将加密方式单独放入一个自定义类文件 LocalPasswordEncoder.java 中:

```java
package com.zioer;

public class LocalPasswordEncoder implements PasswordEncoder{

    @Override
    public String encode(CharSequence rawPassword) {
        return new BCryptPasswordEncoder().encode(rawPassword);
    }

    @Override
    public boolean matches(CharSequence rawPassword, String encodedPassword) {
        BCryptPasswordEncoder encode = new BCryptPasswordEncoder();
        return encode.matches(rawPassword,encodedPassword);
    }
}
```

以上代码中,类 LocalPasswordEncoder 实现了接口 PasswordEncoder,覆盖重写其中两个重要方法 encode 和 matches,encode 用于加密,matches 用于匹配。

更改配置文件 SecurityConfiguration.java 中一个方法,如下代码所示:

```java
package com.zioer;

@Configuration
@EnableWebSecurity
public class SecurityConfiguration extends WebSecurityConfigurerAdapter {

    @Override
    protected void configure(AuthenticationManagerBuilder auth)
        throws Exception {
        auth.inMemoryAuthentication().passwordEncoder(new LocalPasswordEncoder())
            .withUser("user")
            .password(new LocalPasswordEncoder().encode("user"))
            .roles("USER")
            .and()
            .withUser("admin")
            .password(new LocalPasswordEncoder().encode("admin"))
            .roles("USER", "ADMIN");

    }
    ...
}
```

以上代码中,将加密方式替换为前面自定义类 LocalPasswordEncoder。那么,如果需要替换加密方式,则更改类 LocalPasswordEncoder 即可,如以下代码所示:

```
package com.zioer;

public class LocalPasswordEncoder implements PasswordEncoder{

    @Override
    public String encode(CharSequence rawPassword) {
        return rawPassword.toString();
    }

    @Override
    public boolean matches(CharSequence rawPassword, String encodedPassword) {
        return encodedPassword.equals(rawPassword.toString());
    }
}
```

上面代码中,更换了加密方式,实际上只是存储原始密码,密码对比方式为字符串对比。但这种方式,运行代码不会报错。目的是,通过将加密方式自定义类化或接口化,使程序开发更具有灵活性,只需更改加密方式所在类即可。

11.5 使用 UserDetailsService

前面介绍 Spring Boot 2 中集成 Spring Security,以及简单使用方法,方法比较简单。即将用户名和密码直接写在代码中,不利于程序开发的灵活性。基于此,Spring Security 提供了两种方式来灵活获取和验证用户信息。本节介绍其中一种方式:UserDetailsService。

UserDetailsService 是 Spring Security 提供的一个接口,只有一个方法:

```
UserDetails loadUserByUsername(String username);
```

该方法根据输入的用户名查找用户信息,并返回。具体实现由开发人员编写和实现。下面创建一个自定义类 UserDetailsServiceImpl,代码如下所示:

```
package com.zioer.service.impl;

@Service
public class UserDetailsServiceImpl implements UserDetailsService {

    @Override
    public UserDetails loadUserByUsername(String username) throws UsernameNotFoundException {

        if(username.equals("user")) {
            PasswordEncoder encoder = PasswordEncoderFactories
                .createDelegatingPasswordEncoder();
            String password = encoder.encode("user");

            UserDetails user = User.withUsername("user")
                .password(password)
```

```
                .roles("USER")
                .build();

            return user;
        } else {
            throw new UsernameNotFoundException(username);
        }
    }
}
```

上面代码中,类 UserDetailsServiceImpl 实现了接口 UserDetailsService,其中只有一个方法 loadUserByUsername。该方法中,只判断输入是否为用户名 user,并返回一个用户信息 UserDetails。

接着,改写配置类 SecurityConfig 中方法 configure,如下所示:

```
package com.zioer.config;

import com.zioer.service.impl.UserDetailsServiceImpl;

@Configuration
@EnableWebSecurity(debug = true)
public class SecurityConfig extends WebSecurityConfigurerAdapter {

    @Autowired
    UserDetailsService userDetailsService;

    @Bean
    public UserDetailsService createUserDetailsService() {
        return new UserDetailsServiceImpl();
    }

    @Override
    public void configure(AuthenticationManagerBuilder auth) throws Exception {
        PasswordEncoder encoder = PasswordEncoderFactories
            .createDelegatingPasswordEncoder();
        auth
            .userDetailsService(userDetailsService)
            .passwordEncoder(encoder);
    }
    ...
}
```

在上面代码中,注解@EnableWebSecurity 加入属性 debug,值为 true,表示当前为调试中,打印输出相关内容,便于查看。在实际部署时,需要将其置为 false,或删除即可。

重启服务,在弹出登录输入框中,输入用户名 user 和密码 user,可登录系统,输入其他值将没有任何反应。

通过以上简单示例,以实现接口 UserDetailsService 的认证方式,具有更大灵活性。可以在该接口中,采用任何一种方式认证用户,比如可以将用户认证信息放入 H2、MariaDB、MongoDB 等数据库中。

下面实现一个稍微复杂的认证方式,即将用户名和密码存入 H2 数据库,在实现接口 UserDetailsService 中采用 JPA 方式进行认证。编辑 pom.xml 文件,加入相关依赖,示例代码如下所示:

```xml
<dependencies>
    <dependency>
        <groupId>org.springframework.boot</groupId>
        <artifactId>spring-boot-starter-data-jpa</artifactId>
    </dependency>
    <dependency>
        <groupId>org.springframework.boot</groupId>
        <artifactId>spring-boot-starter-security</artifactId>
    </dependency>
    <dependency>
        <groupId>org.springframework.boot</groupId>
        <artifactId>spring-boot-starter-web</artifactId>
    </dependency>
    <dependency>
        <groupId>com.h2database</groupId>
        <artifactId>h2</artifactId>
    </dependency>
</dependencies>
```

上面依赖中,加入 JPA、Security、Web 和 H2 驱动相关依赖。编写 Model 层,建立模型 User 和 Role,User 类代码如下所示:

```java
package com.zioer.model;

@Entity
public class User implements UserDetails {

    @Id
    @GeneratedValue(
            strategy = GenerationType.AUTO,
            generator = "native"
    )
    @GenericGenerator(
            name = "native",
            strategy = "native"
    )
    private Long id;

    @Column(nullable = false, unique = true)
    private String username;
    private String password;
    private boolean accountNonExpired;
    private boolean accountNonLocked;
    private boolean credentialsNonExpired;
    private boolean enabled;
```

```
        @Enumerated(EnumType.STRING)
        @ElementCollection(fetch = FetchType.EAGER)
        private List<Role> roles;

        @Override
        public Collection<? extends GrantedAuthority> getAuthorities() {
            List<GrantedAuthority> authorities = new ArrayList<>();
            roles.forEach(role -> authorities.add(new
              SimpleGrantedAuthority(role.toString())));
            return authorities;
        }

        @Override
        public boolean isAccountNonExpired() {
            return accountNonExpired;
        }

        @Override
        public boolean isAccountNonLocked() {
            return accountNonLocked;
        }

        @Override
        public boolean isCredentialsNonExpired() {
            return credentialsNonExpired;
        }

        @Override
        public boolean isEnabled() {
            return enabled;
        }
        //省略 get 和 set 方法
    }
```

在上面代码中,类 User 实现接口 UserDetails 及其中几个方法。创建类 Role,示例代码如下:

```
package com.zioer.model;

public enum Role {
    USER,
    ADMIN
}
```

以上代码中,为了示例简单,Role 定义为枚举,在实际开发中,应根据需要定义为一个更加通用的类,将角色信息存入数据表中。

下面创建资源类 UserRepository,代码如下所示:

```
package com.zioer.repository;

public interface UserRepository extends JpaRepository<User, Long> {
```

```
    Optional<User> findByUsername(String username);
}
```

上面代码中，接口 UserRepository 继承自 JpaRepository<>，直接使用 JpaRepository 提供的方法，并自定义接口 findByUsername()，即通过用户名查找 User。

创建服务层 UserDetailsServiceImp，示例代码如下所示：

```
package com.zioer.service.impl;

@Service
public class UserDetailsServiceImp implements UserDetailsService {
    @Autowired
    private UserRepository userRepository;

    @Override
    public UserDetails loadUserByUsername(String username)
        throws UsernameNotFoundException {
        Optional<User> user = userRepository.findByUsername(username);

        if (user.isPresent()){
            return user.get();
        }else{
            throw new UsernameNotFoundException(String.format("Username[%s] not found"));
        }
    }
}
```

在上面代码中，类 UserDetailsServiceImp() 实现接口 UserDetailsService，其中，覆盖重写方法 loadUserByUsername()，该方法通过 JPA 方式，返回用户认证信息。下面创建自定义加密类：

```
package com.zioer.config;

@Configuration
public class BCryptEncoderConfig {

    @Bean
    public BCryptPasswordEncoder passwordEncoder(){
        return new BCryptPasswordEncoder();
    }
}
```

上面代码中，创建配置类 BCryptEncoderConfig，其中方法 passwordEncoder() 采用 @Bean 注解。创建 Security 配置类：

```
package com.zioer.config;

@Configuration
@EnableWebSecurity(debug = true)
public class SecurityConfig extends WebSecurityConfigurerAdapter {
```

```java
@Autowired
private UserDetailsService userDetailsService;

@Autowired
private PasswordEncoder passwordEncoder;

@Override
protected void configure(AuthenticationManagerBuilder auth) throws Exception {
    auth.authenticationProvider(authenticationProvider());
}

@Override
protected void configure(HttpSecurity http) throws Exception {
    http
        .httpBasic()
        .and()
        .authorizeRequests()
        .anyRequest().authenticated();
}

@Bean
public DaoAuthenticationProvider authenticationProvider(){
    DaoAuthenticationProvider provider = new DaoAuthenticationProvider();
    provider.setPasswordEncoder(passwordEncoder);
    provider.setUserDetailsService(userDetailsService);
    return provider;
}
}
```

在以上代码中,重点部分是方法 authenticationProvider(),其中,需要设置用于编码和验证密码的 PasswordEncoder 实例,以及设置 UserDetailsService。作为验证,下面创建 Controller 类:

```java
package com.zioer.controller;

@RestController
@RequestMapping(value = "/api/users")
public class UserController {
    @Autowired
    private UserRepository userRepository;

    @GetMapping()
    public List<User> getAllUsers() {
        return userRepository.findAll();
    }

    @GetMapping("/{id}")
    public  User getUserById(@PathVariable Long id) {
        Optional<User> user = userRepository.findById(id);
        if (user.isPresent()) {
```

```
            return user.get();
        }else {
            return null;
        }
    }
}
```

以上代码中,创建了两个简单方法,即通过 Rest 方式查询用户数据。更改启动类,加入测试用数据,示例代码如下所示:

```
package com.zioer;

@SpringBootApplication
public class C114Application {

    public static void main(String[] args) {
        SpringApplication.run(C114Application.class, args);
    }

}

@Component
class FirstCommandLineRunner implements CommandLineRunner{

    @Autowired
    private UserRepository userRepository;
    @Autowired
    private PasswordEncoder passwordEncoder;

    @Override
    public void run(String... args) throws Exception {

        User user = new User();
        user.setUsername("user");
        user.setPassword(passwordEncoder.encode("123456"));
        user.grantAuthority(Role.USER);

        userRepository.save(user);

        user = new User();
        user.setUsername("admin");
        user.setPassword(passwordEncoder.encode("admin"));
        user.grantAuthority(Role.ADMIN);

        userRepository.save(user);


    }
}
```

上面代码在启动时,会在数据库中加入两个测试用用户。启动系统,在浏览器中输入下面地址：

http://127.0.0.1:8080/api/users

将弹出登录框，输入正常的用户名和密码后，才会显示数据，如图 11.4 所示。

```
[{"id":1,"username":"user","password":"$2a$10$FDitqY8mlFaG3bAlG4wTNuaTA4cYbeejZzgOtLip5hpTDRvsmF4KW","accountNonExpired":true,"accountNonLocked":true,"credentialsNonExpired":true,"enabled":true,"authorities":[{"authority":"USER"}]},
{"id":2,"username":"admin","password":"$2a$10$28N8Pp6H5BEa/kfEC6n5TeECgt7gghJgxRkCQuuxra/nbWuKXVCPi","accountNonExpired":true,"accountNonLocked":true,"credentialsNonExpired":true,"enabled":true,"authorities":[{"authority":"ADMIN"}]}]
```

图 11.4 登录后显示示例

由图 11.4 可知，用户的密码是经过加密后存储的，有效提高系统安全性。

11.6 使用 JDBC 认证方式

本节介绍 Spring Security 提供 JDBC 认证方式。创建 SQL 数据文件 schema.sql，内容如下所示：

```sql
DROP TABLE IF EXISTS authorities;
DROP TABLE IF EXISTS users;

create table users (
    username varchar(50) not null primary key,
    password varchar(120) not null,
    enabled boolean not null
);

create table authorities (
    username varchar(50) not null,
    authority varchar(50) not null,
    foreign key (username) references users (username)
);
```

以上 SQL 创建了两张数据表，users 表示用户，authorities 表示权限，之间通过 username 字段关联，创建初始数据文件 data.sql，插入初始数据，示例如下所示：

```sql
insert into users(username, password, enabled) values('admin',
'$2a$10$UhmLDfLiaRnbpbPd6NNYOORILNWmIVlXNG6vgX3G65IsomimOLeSK',true);
insert into authorities(username,authority)values('admin','ROLE_ADMIN');
insert into users(username, password, enabled)values('user',
'$2a$10$qJVTU553hcVdsRvBmSS1Be641iSIRCv5ZnCMB9rHER0qKjqW6Rily',true);
insert into authorities(username,authority)values('user','ROLE_USER');
```

以上 SQL 语句用于插入数据，其中 password 字段数据经过加密处理，密码和用户名相同。下面编辑 Security 配置类 SecurityConfiguration，主要代码如下：

```java
package com.zioer.config;

@Configuration
```

```java
@EnableWebSecurity
public class SecurityConfiguration extends WebSecurityConfigurerAdapter {
    @Autowired
    DataSource dataSource;

    @Autowired
    private PasswordEncoder passwordEncoder;

    //Enable jdbc authentication
    @Autowired
    public void configAuthentication(AuthenticationManagerBuilder auth)
throws Exception {
        auth.jdbcAuthentication()
            .dataSource(dataSource)
            .passwordEncoder(passwordEncoder);
    }
    ...
}
```

上面代码中,主要代码是方法 configAuthentication,该方法采用了 JDBC 进行认证,制定了数据源和加密方法。由于本示例采用 H2 内存数据库,以及两个 sql 文件放置在 resources 文件夹中,可以不进行任何配置,项目结构如图 11.5 所示。

启动服务,使用用户登录验证操作。

以上代码很简单,没有编写任何额外代码便完成了配置。即如果定义的用户表和权限表如同前面定义,则可以不写任何多余代码,便能轻易完成认证,但需要注意的是,用户数据表中 password 加密方式需要和认证加密方法相同。

图 11.5 示例项目结构

但如果用户表和密码表定义和上面方法不相同,则 Spring Security 提供了一种方案,更具有灵活性,此时,需要更改前面 configAuthentication 的代码,如下所示:

```java
public void configAuthentication(AuthenticationManagerBuilder auth)
throws Exception {
auth.jdbcAuthentication()
        .dataSource(dataSource)
        .passwordEncoder(passwordEncoder)
        .usersByUsernameQuery("SELECT username, password, enabled FROM users where username = ?")
        .authoritiesByUsernameQuery ( " SELECT username, authority FROM authorities where username = ?");
}
```

在上面代码中,增加方法 usersByUsernameQuery 和 authoritiesByUsernameQuery,分别是按用户名查找用户和按用户名查找用户权限。由此可知,采用 JDBC 认证方式,提供了一种更加灵活的方式,比如一个老的系统升级改造,可以采用这种方式。

而且当用户名和密码等不是以上命名了,本方法仍然可行。可以尝试更改一下 password 定义为 password2,同样能访问。但需要注意的是,查询顺序不能发生改变。

11.7 带前端认证

前面各节详细介绍了 Spring Boot 2 中集成 Spring Security 的多种方法，本节将结合前面介绍知识和前端处理方法，详细介绍 Spring Security 权限控制和用户登录。

如果一个系统登录方式采用弹框方式，则体验性不够友好。主流方式是访问系统后，单击登录按钮，打开登录界面，输入用户名和密码后，登录系统，按照权限进行页面操作。

创建一个 Spring Boot 2 项目，编辑 pom.xml 文件，加入相关依赖，如下所示：

```xml
<dependencies>
    <dependency>
        <groupId>org.springframework.boot</groupId>
        <artifactId>spring-boot-starter-jdbc</artifactId>
    </dependency>
    <dependency>
        <groupId>org.springframework.boot</groupId>
        <artifactId>spring-boot-starter-security</artifactId>
    </dependency>
    <dependency>
        <groupId>org.springframework.boot</groupId>
        <artifactId>spring-boot-starter-web</artifactId>
    </dependency>
    <dependency>
        <groupId>com.h2database</groupId>
        <artifactId>h2</artifactId>
        <scope>runtime</scope>
    </dependency>
    <dependency>
        <groupId>org.springframework.boot</groupId>
        <artifactId>spring-boot-starter-thymeleaf</artifactId>
    </dependency>
    <!-- Optional, for bootstrap -->
    <dependency>
        <groupId>org.webjars</groupId>
        <artifactId>bootstrap</artifactId>
        <version>4.3.1</version>
    </dependency>
</dependencies>
```

在以上代码中，使用 H2 数据库，加入 Thymeleaf 前端模板以及 bootstrap，用于美观前端页面。编辑 Security 配置文件，主要代码如下所示：

```java
package com.zioer.config;

@Configuration
@EnableWebSecurity
public class SecurityConfiguration extends WebSecurityConfigurerAdapter {
    @Autowired
```

```java
    DataSource dataSource;

    @Autowired
    private PasswordEncoder passwordEncoder;

    //Enable jdbc authentication
    @Autowired
    public void configAuthentication(AuthenticationManagerBuilder auth)
            throws Exception {
        auth.jdbcAuthentication()
        .dataSource(dataSource)
        .passwordEncoder(passwordEncoder);
    }

    @Override
    public void configure(WebSecurity web) throws Exception {
        web.ignoring().antMatchers("/static/**");
        web.ignoring().antMatchers("/webjars/**");
    }

    @Override
    protected void configure(HttpSecurity http) throws Exception {
        http.authorizeRequests()
        .antMatchers("/")
        .permitAll()
        //角色 USER 和 ADMIN 都能访问/list
        .antMatchers("/list").hasAnyRole("USER", "ADMIN")
        //只有角色 ADMIN 能访问/add
        .antMatchers("/add").hasAnyRole("ADMIN")
        .anyRequest().authenticated()
        .and()
        //登录页面,所有人都能访问
        .formLogin().loginPage("/login").permitAll()
        .and()
        //登出页面,所有人都能访问
        .logout().permitAll()
        .logoutSuccessUrl("/");   //退出后访问

        http.csrf().disable();
    }
}
```

以上代码中,web.ignoring().antMatchers()配置忽略指定文件或目录中内容,便于用户在没有登录时,也能进行访问,比如 css、js 文件等。在 configure()方法中,配置角色控制,重要内容已注释在代码上。通过以上配置,明确不同权限用户能访问的内容。

下面创建一个简单的控制类 MainController,示例代码如下:

```java
package com.zioer.controller;

@Controller
```

```java
public class MainController {

    @GetMapping(value = "")
    public String home(Model model) {
        model.addAttribute("msg", "欢迎访问系统首页.");

        return "home";
    }

    @GetMapping(value = "/login")
    public String login(Model model, String error, String logout) {
        if (error != null)
            model.addAttribute("errorMsg", "用户名或密码错误.");

        if (logout != null)
            model.addAttribute("msg", "登出成功.");

        return "login";
    }

    @GetMapping(value = "/list")
    public String list(Model model) {
        model.addAttribute("msg", "列表页面.");

        return "list";
    }

    @GetMapping(value = "/add")
    public String add(Model model) {
        model.addAttribute("msg", "新增页面.");

        return "add";
    }
}
```

根据以上代码中各方法返回值,在目录 resources/templates 中创建对应的页面,同时,创建一个权限错误展示页面 error.html。以上示例代码的项目结构如图 11.6 所示。

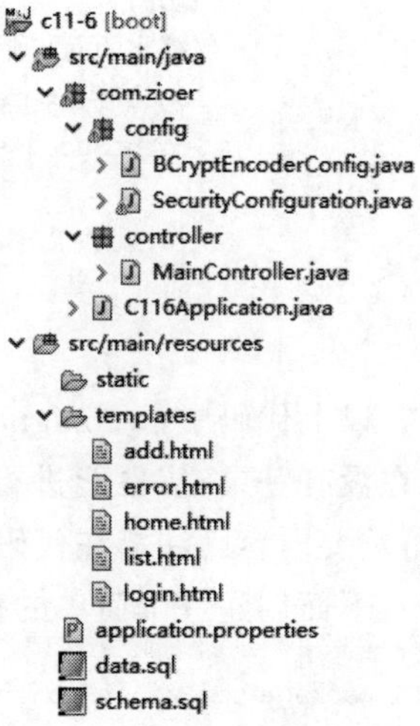

其他页面的代码详见源码。运行本工程,浏览器中访问如下地址:

　　　　http://127.0.0.1:8080/

首页如图 11.7 所示。

首页默认任何用户都可访问,单击登录、列表或新增按钮时,都会打开登录页面,如图 11.8 所示。

该登录页面可以根据系统整体布局进行自定义设计,增加友好性。此时可以使用具有不同权限的用户登录系统进行权限测试,比如具有 USER 权限的用户不能访问需要

图 11.6　示例代码结构

图 11.7 示例首页

图 11.8 登录页面

ADMIN 权限访问的新增页面。

本章小结

本章详细介绍了 Spring Boot 2 中如何引入和使用 Spring Security 安全框架。接着,结合具体案例逐步深入介绍了 Spring Security 的多种配置方式,并结合前端技术,介绍 Spring Security 控制前端页面。本章提供各节详细源码。

第12章 Spring Boot 2测试

本章将介绍 Spring Boot 2 中测试相关技术 JUnit 5。

12.1 JUnit 5 框架介绍

测试是代码开发中重要的一部分。开发人员编写功能模块代码,及时运行代码,调试程序是否运行成功,这就是测试。测试能保证代码的交互质量,及系统是否能稳定运行。发展至今,测试已划分很细,包括但不限于下面几类:

- 单元测试:又称模块测试,即对程序模块进行的测试;
- 集成测试:又称组装测试,在单元测试的基础上,对模块集成后进行测试;
- 系统测试:即将整个软件看作一个整体来进行测试;
- 验收测试:交付前的测试。

如果按是否查看源代码来划分,则可分为:

- 黑盒测试:即不考虑代码内部结构和特性,只注重软件的功能需求的测试;
- 白盒测试:即研究程序结构和源代码的测试。

由此,可知测试在软件项目开发中占据的重要性,从代码编写第一行开始,直至最后交付给用户,都离不开测试。特别是在一个团队中,代码测试更是占据大部分工作,一是需要保证每个开发人员代码的正确性,其次,保证代码整合后的正确性。

测试工具比较多,大致可分为前端测试工具和后端测试工具,前端测试工具重在用户体验、功能完整性测试,减轻用户手动测试的烦琐性,目前有很多用于前端测试的工具,包括 Postman、Selenium、QUnit 等,后端测试工具包括 JUnit 等。

本章介绍的 JUnit 是用于编写和运行可重复的自动化测试的开源测试框架,以保证代码按预期运行。而 JUnit 5 是基于 Java 8 及以上运行,并用到了许多 Java 8 的特性,比如 Lambda 表达式和接口默认方法等。

JUnit 可用于测试对象的一部分方法或一些方法,以及对象之间的互动。其具有的特点有:

- 编写和运行测试的开源框架；
- 提供了注释，以确定测试方法；
- 提供断言，测试预期结果；
- 可以帮助更快地编写代码，提高质量；
- 简单易学、不复杂，不需要花费太多的时间；
- 可自动运行，并提供即时反馈；
- 可以组织成测试套件包含测试案例；
- 能显示测试进度，如果测试没有问题，条形是绿色的，测试失败则会变成红色。

JUnit 5 不同于之前版本只包含在一个 jar 包中，包含 3 个 jar 包，如下所示：

- JUnit Platform：定义了 TestEngine API，用于开发在平台上运行的新测试框架。同时，提供了一个 Console Launcher，用于从命令行启动平台并为 Gradle 和 Maven 构建插件；
- JUnit Jupiter：其包括用于编写测试的新编程和扩展模型。它具有所有新的 JUnit 注解和 TestEngine 的实现，得以运行使用这些注解编写的测试；
- JUnit Vintage：其主要目的是支持在 JUnit 5 平台上运行 JUnit 3 和 JUnit 4 编写的测试。

目前，主流 Java 开发 IDE 都集成了 JUnit，比如 Eclipse、Idea 等，可见 JUnit 在 Java 代码开发中具有重要性。

12.2　Spring Boot 2 集成 JUnit 5

为了更直观了解测试框架在 Spring Boot 2 中的应用，本节将基于前面章节的知识，直接在项目中集成 JUnit 5 测试框架。

目前，应用很广泛的 JUnit 框架版本是 4.x，但 JUnit 5 提供了更广泛的特性，但只支持 Java 8 以上，故很多还是 Java 7 以下的应用只能使用 4.x 以下版本。在 Spring Boot 2 的项目中加入下面依赖：

```xml
<dependencies>
    <dependency>
        <groupId>org.springframework.boot</groupId>
        <artifactId>spring-boot-starter-web</artifactId>
    </dependency>
    <dependency>
        <groupId>org.springframework.boot</groupId>
        <artifactId>spring-boot-starter-test</artifactId>
        <scope>test</scope>
    </dependency>
</dependencies>
```

默认加入的是 junit-4.12.jar，在本书中介绍的是 JUnit 5 以上版本，则需要排除旧版本，并加入新版本，依赖写法如下所示：

```xml
<dependencies>
    <dependency>
        <groupId>org.springframework.boot</groupId>
        <artifactId>spring-boot-starter-web</artifactId>
    </dependency>

    <dependency>
        <groupId>org.springframework.boot</groupId>
        <artifactId>spring-boot-starter-test</artifactId>
        <scope>test</scope>
        <!-- exclude junit 4 -->
        <exclusions>
            <exclusion>
                <groupId>junit</groupId>
                <artifactId>junit</artifactId>
            </exclusion>
        </exclusions>
    </dependency>

    <!-- junit 5 -->
    <dependency>
        <groupId>org.junit.jupiter</groupId>
        <artifactId>junit-jupiter-api</artifactId>
        <version>5.5.1</version>
    </dependency>
    <dependency>
        <groupId>org.junit.jupiter</groupId>
        <artifactId>junit-jupiter-engine</artifactId>
        <version>5.5.1</version>
    </dependency>
    <dependency>
        <groupId>org.junit.platform</groupId>
        <artifactId>junit-platform-engine</artifactId>
        <version>1.5.1</version>
    </dependency>
    <dependency>
        <groupId>org.junit.platform</groupId>
        <artifactId>junit-platform-commons</artifactId>
        <version>1.5.1</version>
    </dependency>
</dependencies>
```

以上依赖写法，将自动加入 JUnit5 的最新版本，如图 12.1 所示。

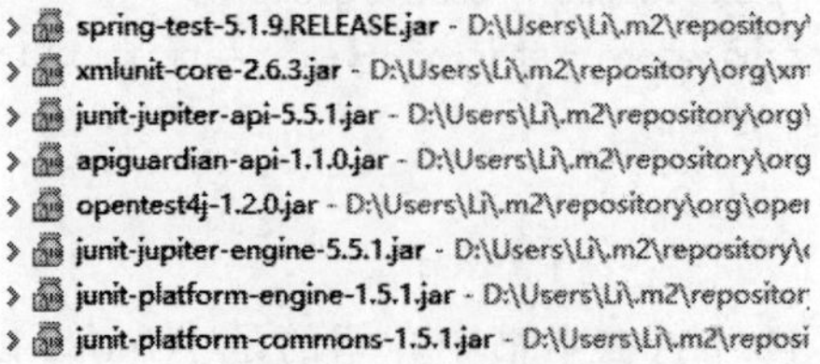

图 12.1　JUnit 5 依赖

由于篇幅限制，图12.1只截取了加入JUnit相关部分jar包。

提示：上面JUnit依赖写法中，加入了version元素，值为5.5.1和1.5.1，这是由于当前Spring Boot 2版本2.2.6.RELEASE默认加入JUnit的版本为5.3.2，在这里，可以为其指定一个最新稳定版本，以使用最新特性。采用以上方法加入后，会导致在pom.xml文件中JUnit相关依赖的版本下面出现警告，如图12.2所示。

图12.2　JUnit相关依赖版本警告

此时，可以将鼠标指针移动到version元素上，在弹出的提示框中，选择Ignore this warning选项即可。

下面是JUnit 5简单测试用例：

```
package com.zioer;

@ExtendWith(SpringExtension.class)
@SpringBootTest(webEnvironment = WebEnvironment.RANDOM_PORT)
public class C121ApplicationTests {
    @LocalServerPort
    int randomServerPort;

    @Autowired
    private TestRestTemplate restTemplate;

    @Test
    @DisplayName("test Hello REST API ")
    public void testHello() {
        String msg = this.restTemplate.getForObject("/hello", String.class);
        assertEquals("Hello World", msg);
    }
}
```

上面代码是一个简单测试类，用到如下注解：

- @ExtendWith：由Jupiter提供的标记接口，在测试类或方法上注册自定义扩展的方法，让Jupiter测试引擎调用给定类或方法的自定义扩展。SpringExtension在Spring 5中引入，用于将Spring TestContext与JUnit 5 Jupiter Test集成；
- @SpringBootTest：由Spring Boot提供的一种方便方法，启动要在测试中使用的应用程序上下文；
- @Test：告诉JUnit附加的公共void方法可以作为测试用例运行；
- @DisplayName：用于为带注释的测试类或测试方法声明自定义显示名称。

创建一个简单 Rest Controller,示例如下所示:

```
package com.zioer.controller;

@RestController
public class HelloController {
    @GetMapping("/hello")
    public String hello() {
     return "Hello World";
    }
}
```

以上代码中方法 hello()比较简单,只返回一个字符串。运行测试代码的方法是打开测试文件 C121ApplicationTests.java,在该文件工作区中,单击鼠标右键,选择 Run As→JUnit Test 命令,运行正确后,如图 12.3 所示。

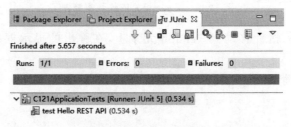

图 12.3　JUnit 运行成功

运行测试的另一种方式是在工程名称上,单击鼠标右键,选择 run as→maven test 命令。

测试运行成功后,将显示绿色进度条。如果 JUnit 运行错误,则会显示红色进度条,如图 12.4 所示。

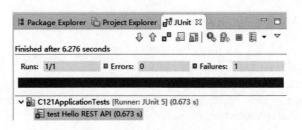

图 12.4　JUnit 运行错误

并会有错误提示窗口,指示开发人员进行修订。JUnit 这种测试方式不同于 Debug 方式,开发人员可以将测试固化,更改代码后,重新运行测试代码,便可将所有可能的测试路径都进行测试,简化开发人员重复工作,有效提高工作效率。

以上测试方式是模拟浏览器打开方式进行测试,所以在测试时,整个工程是按实际工作方式启动服务并运行一遍。Spring 提供了另一种测试方式,即 MockMvc 模式,这种模式无需启动服务,示例代码如下所示:

```
package com.zioer;
```

```
@SpringBootTest
@AutoConfigureMockMvc
public class DemoMockMvcTest {
    @Autowired
    private MockMvc mvc;

    @Test
    public void test() throws Exception {
        RequestBuilder request;

        //使用 get 查一下 hello Controller
        request = get("/hello");
        mvc.perform(request)
                .andExpect(status().isOk())
                .andExpect(content().string(equalTo("Hello World")));
    }
}
```

以上代码用到注解@AutoConfigureMockMvc,表示自动注入 MockMvc,类 MockMvc 是服务器端 Spring MVC 测试支持的主要入口点,其不会加载整个 Spring 容器,模拟了对 Http 的请求,将其转换为对 Controller 的调用,提高测试效率,并提供一套验证工具,方便和简化了测试过程,避免了启动速度慢、依赖网络等问题。

12.3　JUnit 5 使用介绍

12.2 节中,介绍了 Spring Boot 2 中快速集成 JUnit 5 框架,并介绍了两种启动测试方法。尽管测试内容很简单,但由此可见其简便和可重复性,可以很大程度上替代人工重复测试工作,特别是修改代码后,验证其对整个软件工程是否造成破坏时,只需要完整运行一遍测试代码即可。

本节介绍 JUnit 5 的基本使用方法,尤其是新的注解和方法。下面是 JUnit 5 提供的注解:

(1) @TestFactory:动态测试的测试工厂。下面是示例:

```
@TestFactory
Iterator<DynamicTest> dynamicTestsFromCollection() {
    return Arrays.asList(
        DynamicTest.dynamicTest("加法测试", () -> assertEquals(4,
            Math.addExact(2, 2))),
        DynamicTest.dynamicTest("乘法测试", () -> assertEquals(16,
            Math.multiplyExact(4, 4)))
    ).iterator();
}
```

(2) @DisplayName:自定义显示名称。该注解可以放在测试类名或其中方法名之上。下面是示例:

```java
@DisplayName("测试类名")
class TestingAStackDemo {
    @Test
    @DisplayName("测试为空")
    void isEmpty() {
        assertTrue(true);
    }
}
```

该注解将在JUnit标签页中显示定义的名称，特别是测试类和方法较多时，显示名称将更利于理解。

（3）@Nested：表示带注释的类是嵌套的非静态测试类。只有非静态嵌套类（即内部类）才能用作@Nested测试类。嵌套可以任意深入，并且这些内部类被认为是测试类系列的完整成员。下面是示例：

```java
@SpringBootTest
@DisplayName("嵌套测试类")
class NestedTest {
    @Test
    @DisplayName("外部测试类")
    void isEmpty() {
        System.out.println("外部测试类");
        Assertions.assertTrue(true);
    }
    @Nested
    @DisplayName("嵌套一层类")
    class firstNested {
        @Test
        @DisplayName("嵌套一层类")
        void isEmpty() {
            System.out.println("嵌套一层类");
        }
        @Nested
        @DisplayName("嵌套二层类")
        class AfterPushing {
            @Test
            @DisplayName("嵌套二层类")
            void isEmpty() {
                System.out.println("内部二层测试类");
            }
        }
    }
}
```

以上代码示例多层嵌套测试类，在编写复杂测试实例时，很有用。

（4）@Tag：注解用于过滤测试的标签。该注解可用于测试类或方法上，示例如下：

```java
@Tag("Tag测试")
public class TaggedTest {
    @Test
```

```
    @Tag("方法1")
    void testMethod() {
        assertEquals(3 * 2, 6);
    }
}
```

(5) @ExtendWith：用于注册自定义扩展。示例如下：

```
@ExtendWith(SpringExtension.class)
@ContextConfiguration(classes = AppConfig.class)
public class MyAppTest {
}
```

开发人员可对测试接口、类、方法或自定义进行组合注解。

(6) @BeforeEach：表示在每个测试方法之前执行带注释的方法（之前为@Before）。示例如下：

```
@BeforeEach
void init() {
    System.out.println("@BeforeEach - 在每个方法运行前执行");
}
```

(7) @AfterEach：表示在每个测试方法之后执行带注释的方法（之前为@After）。示例如下：

```
@AfterEach
void tearDown() {
    System.out.println("@AfterEach - 在每个方法执行后执行");
}
```

(8) @BeforeAll：表示将在当前类中的所有测试方法之前执行带注释的方法（之前为@BeforeClass）。示例如下：

```
@BeforeAll
static void setup() {
    System.out.println("@BeforeAll - 在所有方法运行前执行一次");
}
```

(9) @AfterAll：表示将在当前类中的所有测试方法之后执行带注释的方法（之前为@AfterClass）。示例如下：

```
@AfterAll
static void done() {
    System.out.println("@AfterAll - 在所有方法运行后前执行");
}
```

(10) @Disable：用于禁用测试类或方法（之前为@Ignore）。示例如下：

```
@Test
@Disabled("Disabled")
@DisplayName("测试为空")
void isEmpty() {
```

```
        assertTrue(true);
}
```

可以在该注解后加入说明性文字,以更好帮助理解。

(11) @TestTemplate:表示方法是测试用例的模板。

(12) @RepeatedTest:表示方法是重复测试的测试模板。示例如下:

```
@DisplayName("加法 test")
@RepeatedTest(5)
void addNumber() {
    Assertions.assertEquals(3, 1 + 2, "1 + 2 should equal 3");
}
```

以上代码中,方法 addNumber 将运行 5 次。

(13) @TestMethodOrder:用于配置测试类中测试方法执行顺序。示例如下:

```
@SpringBootTest
@TestMethodOrder(Alphanumeric.class)
class MethodOrderTest {
    //省略测试方法
}
```

以上代码将按照测试方法的首字母顺序执行。该注解的赋值有如下四种:

① Alphanumeric.class:按照方法字符排序执行;

② OrderAnnotation.class:按照指定注解@order 顺序执行,这种方法灵活性强;

③ Random.class:按照随机模式执行;

④ 最后一种方式是实现接口 MethodOrderer,自定义顺序。

(14) @TestInstance:用于为带注释的测试类配置测试实例生命周期。

(15) @DisplayNameGeneration:声明测试类的自定义显示名称生成器。示例如下:

```
@DisplayNameGeneration(DisplayNameGenerator.ReplaceUnderscores.class)
class DisplayNameGenerationTest {
    @Test
    void test_generator() {
    }
    @Test
    void do_something_should_happen() {
    }
}
```

在上面代码中,使用该注解指定生成显示名称为替换下画线为空格。

(16) @TempDir:用于通过生命周期方法或测试方法中的字段注入或参数注入来提供临时目录。

以上是 JUnit Jupiter 支持的注解,用于配置测试和扩展框架。JUnit 5 继承许多 JUnit 4 的断言方法,并增加适合与 Java 8 lambdas 一起使用的方法。下面是常用断言:

(1) assertEquals():用于断言期望值和实际值相等,并为不同的数据类型提供了许多重载方法,例如 int、short 等。示例如下:

```
assertEquals(12, 3 * 4);
```

（2）assertNotEquals()：用于断言期望值和实际值不相等。示例如下：

```
assertNotEquals(9, 3 * 4);
```

（3）assertArrayEquals()：用于断言预期和实际数组是相等的。示例如下：

```
char[] expected = {'z', 'i', 'o', 'e', 'r'};
char[] actual = "zioer".toCharArray();
assertArrayEquals(expected, actual, "Arrays 应该相等");
```

（4）assertIterableEquals()：用于断言预期和实际的迭代是完全相同的，即迭代器必须以相同的顺序返回相同的元素。示例如下：

```
Iterable<String> a = new ArrayList<>(asList("abc", "def"));
Iterable<String> b = new LinkedList<>(asList("abc","def" ));
assertIterableEquals(a, b);
```

（5）assertLinesMatch()：用于断言字符串的预期列表与实际列表匹配。

```
List<String> a = asList("a", "cf");
List<String> b = asList("a","cf");
assertLinesMatch(a, b);
```

（6）assertNull()：用于断言实际为 null。示例如下：

```
List<String> a = null;
assertNull(a);
```

（7）assertNotNull()：用于断言实际为非 null。示例如下：

```
String a = "a string";
assertNotNull(a);
```

（8）assertSame()：用于断言期望和实际引用完全相同的对象。示例如下：

```
String a = "abcde";
Optional<String> b = Optional.of(a);
assertSame(a, b.get());
```

（9）assertNotSame()：用于断言期望和实际没引用同一个对象。

（10）assertTimeout()：用于测试长时间运行的任务。如果 testcase 中的给定任务需要超过指定的持续时间,那么测试将失败。示例如下：

```
assertTimeout(Duration.ofSeconds(2), () -> Thread.sleep(1000));
```

（11）assertTrue()：用于断言所提供的条件为 true。示例如下：

```
assertTrue(3 > 1);
assertTrue(null == null, "null is equal to null");
```

（12）assertFalse()：用于断言所提供的条件为 false。示例如下：

```
assertFalse(1 > 3);
```

(13) assertAll()：该断言允许创建分组断言，所有断言都被执行并且它们的错误一起报告。示例如下：

```
assertAll(
  "测试多个",
  () -> assertEquals(12, 3 * 4),
  () -> assertEquals("zioer", "Zioer".toLowerCase())
);
```

(14) assertThrows()：该断言允许一个明确而简单的方法来断言可执行文件是否抛出指定的异常类型。

(15) fail()：该方法指示没有通过测试。在维护遗留代码时可能很有用。示例如下：

```
@Test
void test() {
    fail("Not yet implemented");
}
```

以上介绍的是在JUnit 5环境下，为了有效利用该框架而提供的多种方法，能提高测试效率，加快测试进度。

在一个项目中，不可能只存在一个测试文件，其弊端是不利于测试的归类、管理和运行。在JUnit 5下，可以快速创建一个测试套件，同时运行不同测试类和测试包。下面是两个新的注解：

- @SelectPackages：用于选择在运行测试套件时的包；
- @SelectClasses：用于选择在运行测试套件时的类。

创建一个测试套件，指定多个包的方式示例，如下所示：

```
package com;

@RunWith(JUnitPlatform.class)
@SelectPackages("com.zioer")
public class AllTestsSelectPackages {
}
```

以上代码使用注解@SelectPackages指定单个运行包，当运行该套件类时，则会执行指定包下所有测试文件。

当要同时指定多个运行包时，则需要将包组合为字符串数组传递给注解@SelectPackages，示例如下所示：

```
package com;

@RunWith(JUnitPlatform.class)
@SelectPackages({"com.zioer","com.zioer.a","com.zioer.b"})
public class AllTestsSelectPackages {
}
```

在以上代码中，注解@SelectPackages加入了三个运行包。

如果不希望同时运行一个包中所有测试类，则可以使用注解@SelectClasses组合多个测试类，示例如下所示：

```
package com;

@RunWith(JUnitPlatform.class)
@SelectClasses(NestedTest.class)
public class AllTestsSelectPackages {
}
```

以上代码中，注解@SelectClasses 指定了单个测试类。如果需要同时运行多个测试类时，则需要将多个类组合为一个数组，传递给注解@SelectClasses，示例如下所示：

```
package com;

@RunWith(JUnitPlatform.class)
@SelectClasses({NestedTest.class,RepeatedUnitTest.class})
public class AllTestsSelectPackages {
}
```

本节详细介绍了 JUnit 5 的常用的注解，以及基本使用方法，方便开发人员理解和掌握 JUnit 5 的操作。本节介绍的示例提供完整源码。

12.4　JUnit 5 完整示例

本节将结合前面章节介绍的 Spring Boot 2 操作和本章知识，介绍 JUnit 5 在实际项目中如何进行单元测试，以保证开发程序的正确性。

在开发 Spring Boot 2 RESTful 应用中，很重要部分是对接口的测试，那么 JUnit 5 在接口测试中具有的作用是快速对所有接口进行测试，特别是在对代码修改后，查看其是否对所有接口造成影响，以及在团队开发中，能保证合并代码的质量。

创建一个 Spring Boot 2 的项目，在采用 Spring Tool Suit 4 工具自动生成的结果中，自动生成目录/src/test/java，即建议用于存放测试文件。工程目录结构如图 12.5 所示。

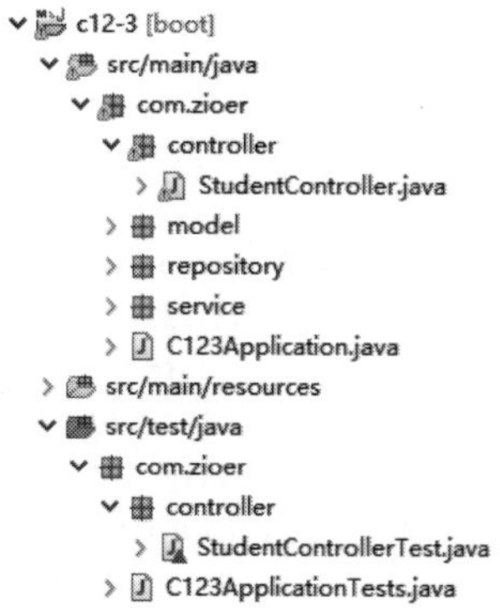

这是一个典型的 RESTful 目录结构，最后的测试文件在/src/test/java/controller 下，其中 C123ApplicationTests.java 是项目自动生成的测试文件。该项目的 pom.xml 文件中依赖项如下所示：

```
<dependencies>
    <dependency>
        <groupId>org.springframework.boot</groupId>
        <artifactId>spring-boot-starter-data-jpa</artifactId>
    </dependency>
    <dependency>
        <groupId>org.springframework.boot</groupId>
```

图 12.5　工程目录

```xml
        <artifactId>spring-boot-starter-web</artifactId>
    </dependency>
    <dependency>
        <groupId>com.h2database</groupId>
        <artifactId>h2</artifactId>
        <scope>runtime</scope>
    </dependency>

    <dependency>
        <groupId>org.springframework.boot</groupId>
        <artifactId>spring-boot-starter-test</artifactId>
        <!-- exclude junit 4 -->
        <exclusions>
            <exclusion>
                <groupId>junit</groupId>
                <artifactId>junit</artifactId>
            </exclusion>
        </exclusions>
    </dependency>

    <!-- junit 5 -->
    <dependency>
        <groupId>org.junit.jupiter</groupId>
        <artifactId>junit-jupiter-engine</artifactId>
        <version>5.5.1</version><!-- $NO-MVN-MAN-VER$ -->
        <exclusions>
            <exclusion>
                <groupId>org.junit.platform</groupId>
                <artifactId>junit-platform-engine</artifactId>
            </exclusion>
            <exclusion>
                <groupId>org.junit.platform</groupId>
                <artifactId>junit-platform-commons</artifactId>
            </exclusion>
        </exclusions>
    </dependency>
    <dependency>
        <groupId>org.junit.jupiter</groupId>
        <artifactId>junit-jupiter-api</artifactId>
        <version>5.5.1</version><!-- $NO-MVN-MAN-VER$ -->
        <exclusions>
            <exclusion>
                <groupId>org.junit.platform</groupId>
                <artifactId>junit-platform-commons</artifactId>
            </exclusion>
        </exclusions>
    </dependency>
    <dependency>
        <groupId>org.junit.platform</groupId>
        <artifactId>junit-platform-engine</artifactId>
        <version>1.5.1</version><!-- $NO-MVN-MAN-VER$ -->
```

```xml
        </dependency>
        <dependency>
            <groupId>org.junit.platform</groupId>
            <artifactId>junit-platform-commons</artifactId>
            <version>1.5.1</version><!-- $NO-MVN-MAN-VER$ -->
        </dependency>
</dependencies>
```

其中，用到了如下几个关键依赖：

- JPA；
- spring-boot-starter-web；
- H2 数据库；
- spring-boot-starter-test；
- JUnit 5。

创建业务模型、初始数据表和 JPA 过程不再赘述。创建控制类 StudentController 的代码如下所示：

```java
package com.zioer.controller;

@RestController
@RequestMapping(value = "/api/students")
public class StudentController {
    @Autowired
    private StudentServiceImp studentService;

    @GetMapping()
    public List<Student> getAll() {
        return studentService.findAll();
    }

    @GetMapping("/{id}")
    public Student getById(@PathVariable Long id) {
        return studentService.findById(id);
    }

    @PostMapping()
    public ResponseEntity<Student> addOne(@Valid @RequestBody Student student)
    {
        Student saved = studentService.insert(student);

        return new ResponseEntity<Student>(saved, HttpStatus.CREATED);
    }

    @PutMapping(value = "/{id}")
    public ResponseEntity<Student> updateOne (@PathVariable("id") long id,
            @Valid @RequestBody Student student) {
        Student stu = studentService.findById(id);
        if (stu == null) {
            return new ResponseEntity<Student>(HttpStatus.NOT_FOUND);
```

```
            }

            student.setId(id);

            Student saved = studentService.update(student);
            return new ResponseEntity<Student>(saved, HttpStatus.OK);
        }

        @DeleteMapping(value = "/{id}")
        public ResponseEntity<HttpStatus> removeOne (@PathVariable("id") int id)
        {
            Student stu = studentService.findById(id);
            if (stu == null) {
                return new ResponseEntity<HttpStatus>(HttpStatus.NOT_FOUND);
            }

            studentService.delete(stu);

            return new ResponseEntity<HttpStatus>(HttpStatus.ACCEPTED);
        }

}
```

以上代码用到如下注解：
- @GetMapping()：用于获得所有记录，链接为/api/students；
- @GetMapping("/{id}")：用于获得单条记录，链接为/api/students/{id}；
- @PostMapping()：新增一条记录，链接为/api/students，方法为Post；
- @PutMapping(value = "/{id}")：更新一条记录，链接为/api/students/{id}，方法为Put；
- @DeleteMapping(value = "/{id}")：删除一条记录，链接为/api/students/{id}，方法为Delete。

建立测试类测试上面接口是否为预期，快捷方式是在控制类StudentController名称上右击，选择New→Other命令，在弹出窗口中选择JUnit Test Case项后，单击Next按钮，在新打开窗口中，已自动填写完成创建测试类名称、所在目录等默认值，确认无误后，单击Finish完成初始化过程，如图12.6所示。

在图12.6中，可选择复选框和单击Next按钮进行高级设置，在此不再一一介绍。

在新打开的编辑窗口中增加配置项，示例代码如下所示：

```
@SpringBootTest
@AutoConfigureMockMvc
@TestMethodOrder(OrderAnnotation.class)
class StudentControllerTest2 {
    @Autowired
    private MockMvc mvc;


}
```

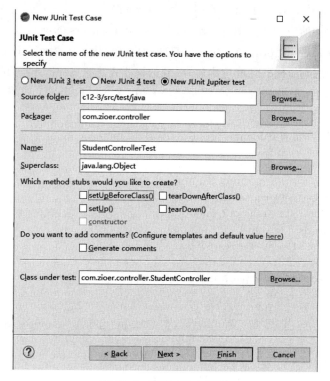

图 12.6　新建 JUnit 测试

在上面代码中，用到如下注解：
- @SpringBootTest：表示下面是一个测试类；
- @AutoConfigureMockMvc：用于测试类，以启用自动配置；
- @TestMethodOrder：设置测试类中方法运行的顺序；
- @Autowired：自动加载和设置 MockMvc。

通过以上几个注解，完成了主要配置工作。下面是测试方法：

```
@Test
@Order(1)
public void testGetList() throws Exception {
    String path = "/api/students";
    // 1. getAll
    mockMvc.perform(MockMvcRequestBuilders
        .get(path)
        .accept(MediaType.APPLICATION_JSON))
        .andDo(print())
        .andExpect(status().isOk())
        .andExpect(content().contentType("application/json;charset=UTF-8"))
        .andExpect(jsonPath("$", hasSize(6)))
        .andExpect(jsonPath("$[0].id").value(1))
        .andExpect(jsonPath("$[1].id").value(2))
        .andExpect(jsonPath("$.[*].id").isNotEmpty());
}
```

以上代码中,方法 testGetList()使用注解@Order(1),表示执行顺序,path 表示测试路径/api/students,返回所有记录。其中,MockMvcRequestBuilders 的 get()方法传递测试路径,MockMvcResultHandlers 的 print()方法用以打印输出响应内容,MockMvcResultMatchers 的多个方法用以验证状态响应代码和响应内容,用到的重要方法有 jsonPath,测试返回 JSON 数据是否符合预期值,比如记录总数、某条记录是否存在等。由于数据库采用 H2,每次运行时,数据都会经过初始化。

下面是测试路径/api/students/{id}返回内容的代码:

```
@Test
@Order(2)
public void testGetOne() throws Exception {
    String path = "/api/students/{id}";
    // 2. getOne
    mockMvc.perform(MockMvcRequestBuilders
            .get(path,1)
            .accept(MediaType.APPLICATION_JSON))
            .andDo(print())
            .andExpect(status().isOk())
            .andExpect(content().contentType("application/json;charset=UTF-8"))
            .andExpect(jsonPath("$.id").exists())
            .andExpect(jsonPath("$.id").value("1"));
}
```

以上代码中,使用了路径变量{id},注意在 get 中 id 的传值方法,实际上最后拼凑路径为/api/students/1。

期望返回为 json 格式数据,并包含值,exists()用以测试是否存在,value("1")表示其值是否为 1。

下面是测试 Post 一条数据新增记录是否正确的方法:

```
@Test
@Order(3)
public void testPostOne() throws Exception {
    String path = "/api/students";
    //3. PostOne
    mockMvc.perform(MockMvcRequestBuilders
            .post(path)
            .content(toJsonString(new Student("萌萌", "女",
                "mengmeng@163.com","13719343254","石家庄",85)))
            .contentType(MediaType.APPLICATION_JSON)
            .accept(MediaType.APPLICATION_JSON))
            .andExpect(status().isCreated())
            .andExpect(jsonPath("$.id").exists())
            .andExpect(jsonPath("$.id").value("7"))
            .andExpect(jsonPath("$.name").value("萌萌"));
}
```

以上代码中,使用了 MockMvcRequestBuilders 中的 post 方法传递路径和待保存的记录值。最后,测试返回内容是否为新增记录。

下面是测试 Put 方法的代码：

```java
@Test
@Order(4)
public void testPutOne() throws Exception {
    String path = "/api/students/{id}";
    //4. PutOne
    mockMvc.perform(MockMvcRequestBuilders
            .put(path, 2)
            .content(toJsonString(new Student("图图", "男",
                "tutu@163.com","13843539806","青岛",68)))
            .contentType(MediaType.APPLICATION_JSON)
            .accept(MediaType.APPLICATION_JSON))
            .andExpect(status().isOk())
            .andExpect(jsonPath("$.id").exists())
            .andExpect(jsonPath("$.id").value("2"))
            .andExpect(jsonPath("$.name").value("图图"))
            .andExpect(jsonPath("$.sex").value("男"));
}
```

在上面代码中，使用了 MockMvcRequestBuilders 中 put 方法，传递了路径和待更新的记录。并测试返回内容是否为修改后的记录。

下面是测试 delete 方法的代码：

```java
@Test
@Order(5)
public void testdeleteOne() throws Exception
{
    String path = "/api/students/{id}";
    //5. DeleteOne
    mockMvc.perform( MockMvcRequestBuilders
            .delete(path, 3))
            .andExpect(status().isAccepted());
}
```

上面代码主要使用了 MockMvcRequestBuilders 的 delete 方法，测试返回状态的正确性。

在上面几个测试方法中，用到自定义方法，目的是将对象转换为 JSON 格式的内容，代码如下所示：

```java
public static String toJsonString(final Object obj) {
    try {
        return new ObjectMapper().writeValueAsString(obj);
    } catch (Exception e) {
        throw new RuntimeException(e);
    }
}
```

以上测试代码保存后，在该文件名上单击右键，选择 Run As→JUnit Test，运行该测试文件，测试结果如图 12.7 所示。

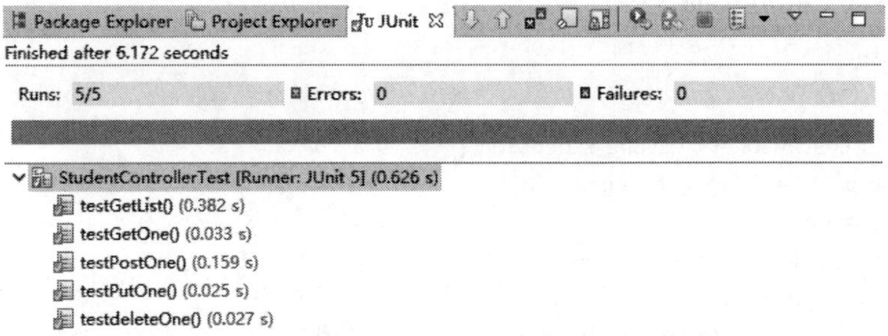

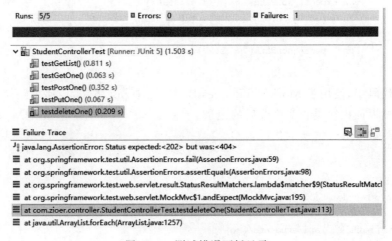

图 12.7　JUnit 测试结果

运行完成后，打开 JUnit 测试面板，显示测试结果和运行时间值。如果在方法 testdeleteOne()中测试删除一条不存在的记录，测试结果如图 12.8 所示。

图 12.8　测试错误面板显示

由图 12.8 所示，提示条变成红色，并有提示哪个测试方法发生错误，并在错误跟踪界面中，能定位找到错误发生行。

本节提供完整示例源码，请查看并运行源码。

12.5　Maven 配置测试环境

在开发中，希望实际运行环境和测试环境能分开，比如在开发测试中，能用到测试数据库 H2，以及纯净的测试数据，等测试完成后再切换到实际运行环境。

一种方式是修改配置文件，在不同的环境间进行切换。这不失为一个方法，缺点是在开发环境切换时，总是会忘记修改配置文件，或是遗漏某些配置，以造成实际数据的污染。

另一种有效方法是编辑项目 pom.xml 文件，配置多个环境，实现快速切换。下面假设配置两个场景，一个是用于测试环境，使用 H2 数据库，另一个用于生产环境，使用 MariaDB 数据库。

创建文件夹 src\main\profile\dev，存放测试环境的配置文件，在其中创建配置文件

application.properties，内容如下所示：

```
server.port = 8082
spring.jpa.show-sql = true
```

测试数据库使用内存数据库 H2，在配置文件中重新指定端口号和设置运行时显示 SQL 语句。

创建文件夹 src\main\profile\local，存放生产环境的配置文件，在其中创建配置文件 application.properties，内容如下所示：

```
spring.datasource.url = jdbc:mysql://localhost:3306/test?useUnicode = true&characterEncoding = UTF-8&useSSL = false&serverTimezone = GMT%2B8
spring.datasource.driver-class-name = org.mariadb.jdbc.Driver
spring.datasource.username = root
spring.datasource.password = root
```

在上面配置文件中，主要是配置 MariaDB 数据库的连接信息，端口采用默认 8080。在以上两个配置文件中，建议 server.port 设置为不同值。

编辑文件 pom.xml，在其中加入 profiles 节点，内容如下所示：

```xml
<profiles>
    <profile>
        <id>local</id>
        <build>
            <resources>
                <resource>
                    <directory>src/main/profile/local</directory>
                </resource>
                <resource>
                    <directory>src/main/resources</directory>
                </resource>
            </resources>
        </build>
    </profile>
    <profile>
        <id>dev</id>
        <activation>
            <activeByDefault>true</activeByDefault>
        </activation>
        <build>
            <resources>
                <resource>
                    <directory>src/main/profile/dev</directory>
                </resource>
                <resource>
                    <directory>src/main/resources</directory>
                </resource>
            </resources>
        </build>
```

```
        </profile>
    </profiles>
```

以上节点 profiles 是关键配置信息,定义两个元素 profile,分别为生产环境和测试环境配置,用元素 id 标识,元素 activeByDefault 用来标识是否为默认环境,在这里将测试环境 dev 标识为默认环境,元素 resources 标识配置文件信息位置。

在项目名称单击右键,选择 Maven→Select Maven profiles 命令,弹出 Select Maven profiles 窗口,如图 12.9 所示。

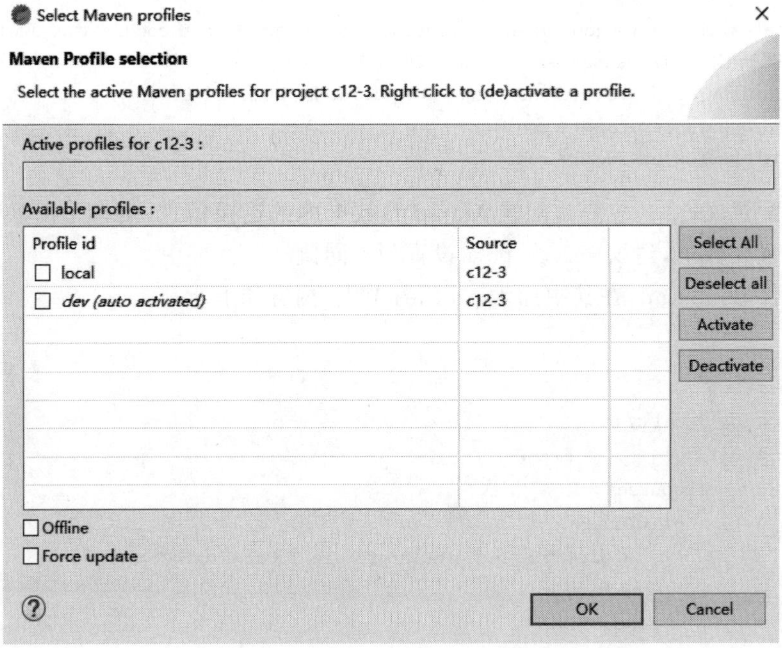

图 12.9　Select Maven profiles 窗口

由图所示,可选择 pom.xml 文件中定义的 profile,默认为 dev。选择 local 项,单击 OK 按钮,再次查看 Project Explorer 面板中该工程,多了 src/main/profile/local 项,如图 12.10 所示。

在图 12.10 界面中,目录 src/main/profile/local 加入了工程源中,即根据用户选择加入 local 目录,其中的配置文件也纳入工程中。同理,当在图 12.9 选择 dev 或不选择时,将在图 12.10 界面中加入目录 src/main/profile/dev。

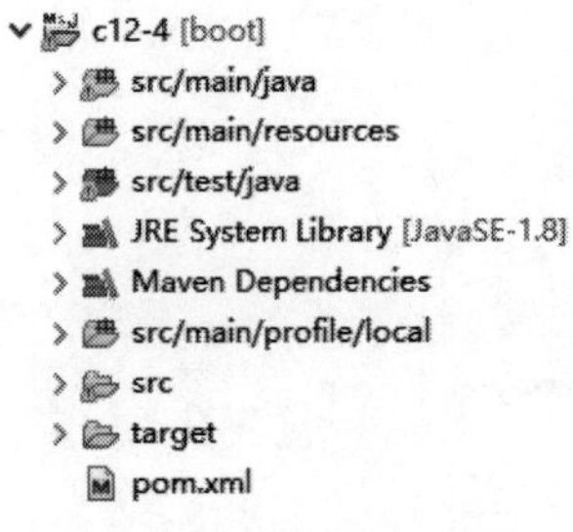

建议在开发和测试中,选择 dev 测试环境。测试完全通过后,再切换为 local 生产环境。以此加快开发进度,并保证软件质量。

尽管在上面介绍中,只在 pom.xml 中加入了两个 profile,实际上,可加入更多个 profile,以适应不同的环境。以上完整代码详见本章提供的源码。

图 12.10　项目结构

本章小结

测试是软件项目开发中重要环节,本章介绍在 Spring Boot 2 中测试相关技术。其中,单元测试是最基本工作,只有严格执行单元测试,软件质量才有可能得到保证,以进行下一步整体测试。同时,本章基于 JUnit 5 的基本操作,介绍了多种测试方法,结合前面章节内容,介绍几个典型案例分析。最后,介绍在开发中测试环境和生产环境间的配置技巧,多环境配置在开发中具有重要作用,面对不同开发人员、不同客户等场景,需要有不同的配置,典型的是分别连接多类型、不同数据库。

第13章 其他相关技术

前面章节结合作者实际开发经验,详细介绍了 Spring Boot 2 中重要知识点,但还是不能面面俱到。本章作为本书最后一章,将集中介绍 Spring Boot 2 中其他相关技术点,同时,它们也是开发中的重要部分,具体包括上传文件、Lombok 应用、Devtools、更改启动 Logo 和缓存等技术。

13.1 上传文件

作为一个在线应用系统,上传文件是其中重要环节,比如让用户上传图片、分享下载等,都离不开上传文件部分。本节介绍在 Spring Boot 2 中如何处理上传文件和在线显示图片技术。

Spring Boot 2 应用的特点是小,一般建议打包为 jar,利于管理和部署。在传统的 Spring MVC 架构中,专门有个存放静态文件和上传文件的地方,如图 13.1 所示。

在 Spring MVC 项目中,使用类似下面代码获得当前上传目录:

图 13.1 Spring MVC 结构

```
ServletContext context = request.getServletContext();      //获取上传文件夹的路径
String uploadPath = context.getRealPath("/upload");        //获取本地存储位置的绝对路径
```

那么在 Spring Boot 2 中就不能采用上面方式获得上传目录。

在建立的 Spring Boot 2 项目中,编辑配置文件 application.properties,加入上传目录等信息,如下所示:

```
spring.resources.static-locations=file:${upload.path},classpath:/static/
spring.static-path-pattern=/**


# Single file max size
```

```
spring.servlet.multipart.max-file-size=10MB
spring.servlet.multipart.max-request-size=10MB

upload.path=d:/images/
```

在上面代码中，upload.path 配置为上传文件的目录；spring.resources.static-locations 表示静态资源访问目录，在这里，需要将项目中目录/static 加上，否则该目录下也不能静态访问，当同时配置多个静态资源目录时，之间以逗号相隔；spring.static-path-pattern 表示访问文件模式，/** 即目录下所有层级文件都可以访问；spring.servlet.multipart.max-file-size 和 spring.servlet.multipart.max-request-size 表示限制上传文件大小。

以上配置很重要，会影响代码控制上传文件。

为上传文件专门创建一个服务类 StorageService，主要代码如下所示：

```
package com.zioer.service;

@Service
public class StorageService {
    @Value("${upload.path}")
    private String path;

    public String uploadFile(MultipartFile file) {
        if (file.isEmpty()) {
            throw new StorageException("没有上传的文件");
        }
        String fileExtension = getFileExtension(file);      //扩展
        String filename = getRandomString();                //新名称

        try {

            File targetFile = getTargetFile(fileExtension, filename);

            file.transferTo(targetFile);

        }catch(IOException e){
            throw new StorageException(e.toString());
        }

        return filename + fileExtension;
    }
    //通过文件名称,返回一个 File 类型
    public File getFile(String filePath) {
        File file = new File(path + filePath);
        return file;

    }
    //获得一个随机名称
    private String getRandomString() {
        return new Random().nextInt(999999) + "_" + System.currentTimeMillis();
    }
```

```java
    //返回一个新文件
    private File getTargetFile(String fileExtn, String fileName) {
        File targetFile = new File(path + fileName + fileExtn);
        return targetFile;
    }
    //获得文件扩展名
    private String getFileExtension(MultipartFile inFile) {
        String fileExtention = inFile.getOriginalFilename().substring(inFile.getOriginalFilename().lastIndexOf('.'));
        return fileExtention;
    }
    //获得文件扩展名
    public String getExName(String fileName) {
        String fileExtention = fileName.substring(fileName.lastIndexOf('.') + 1);
        return fileExtention;
    }
}
```

上面代码详细描述了上传文件的步骤，其中变量 path 通过注解 @Value("${upload.path}") 获得配置文件中定义的上传目录 upload.path，增加了项目应用的灵活性，可在实际部署时只更改配置文件，而无须更改代码层，就达到更改实际上传目录的目的。方法 uploadFile() 是实现上传文件的重要部分，主要是通过获取传递变量 file，新建一个随机名，将传递文件变量 file 保存在上传目录中，并更改为新名称，最后返回存储的名称过程。在其中，用到几个自定义方法，详见代码中注释所示。

上面定义服务类 StorageService 具有通用性，可在多个地方进行调用，下面介绍如何在 RESTful Controller 和单独的 Controller 中调用。创建一个 RESTful Controller 类：

```java
package com.zioer.controller.rest;

@RestController
public class UploadRestController {
    @Autowired
    private StorageService storageService;

    @PostMapping(value = "/api/upload")
    public String handleFileUpload(@RequestBody() @PathVariable MultipartFile file) {
        return storageService.uploadFile(file);
    }
}
```

上面定义的 RestController 类，方法 handleFileUpload 用于处理用户上传文件，代码很简洁，只用一个获得上传文件的变量，调用服务类 StorageService 中 uploadFile 方法，就实现了用户上传文件。

下面创建 Controller 类 UploadController，代码如下所示：

```java
package com.zioer.controller;

@Controller
public class UploadController {
```

```
    @Autowired
    private StorageService storageService;

    @RequestMapping(value = "/upload", method = RequestMethod.POST,
            consumes = {"multipart/form-data"})
    public String upload(@RequestParam MultipartFile file) {

        storageService.uploadFile(file);

        return "redirect:/success.html";
    }
}
```

上面代码中,方法 upload()用于获取用户上传文件,并调用 StorageService 中方法 uploadFile,实现文件的上传。静态 Html 文件在这里不再展示。图 13.2 所示为本项目结构。

图 13.2　项目结构

启动该项目,在 Postman 中输入 RESTful 上传路径:

$$http://localhost:8080/api/upload$$

测试上传文件,如图 13.3 所示。

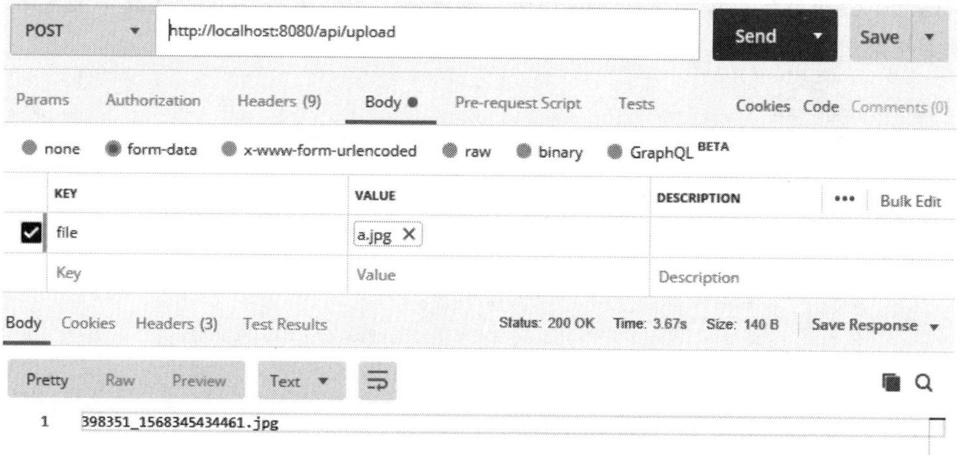

图 13.3　Postman 测试上传文件

如图所示，测试上传结果返回为新建文件名。同理，在浏览器中，输入下面地址：
$$http://127.0.0.1:8080/$$
显示上传文件界面，单击浏览按钮，选择需要上传的文件，如图13.4所示。

图13.4　浏览器上传文件

单击上传按钮，测试上传文件成功。

上传文件大小在前面配置文件中设置为最大10M，如果不进行设置，Spring Boot 2默认文件大小不能超过1M。

提示：前面代码中，没有对上传文件类型进行设置，在实际开发代码中，应该根据需要对上传文件类型进行限制。

在页面显示图片有两种方式。一是通过配置文件进行设置静态文件夹方式，如前面配置文件中介绍：

spring.resources.static-locations=file:${upload.path},classpath:/static/

上面配置中的写法file:${upload.path}，直接引用本页定义的upload.path值，有利于一处定义，处处使用。比如上传后，文件名为398351_1568345434461.jpg，则在浏览器中输入：

$$http://127.0.0.1:8080/398351\_1568345434461.jpg$$

就可以在浏览器页面直接显示该图片。这种显示图片方式比较直接，只要知道图片的地址，便可直接进行访问。

另一种访问方式是，如果不想用静态资源方式访问图片，或不想在配置文件中配置图片访问目录，可以通过代码控制方式进行访问。编辑类UploadController，在其中加入下面方法，用来显示图片：

```
@GetMapping("get")
public void get(String imagepath, HttpServletResponse response) {
    OutputStream stream = null;
    FileInputStream inputStream = null;

    try {
        File file = storageService.getFile(imagepath);
        String type = "image/" + storageService.getExName(imagepath)
            + ";charset=utf-8";

        inputStream = new FileInputStream(file);
        byte[] data = new byte[(int) file.length()];
        inputStream.read(data);
```

```
            response.setContentType(type);
            stream = response.getOutputStream();
            stream.write(data);
            stream.flush();
        } catch(Exception e) {
            e.printStackTrace();
        } finally {
            try {
                inputStream.close();
                stream.close();
            } catch (IOException e) {
                e.printStackTrace();
            }
        }
    }
```

上面代码中，通过传递图片名称，获得服务器上图片的流，将该流推送到页面进行显示，代码不算复杂。在浏览器显示图片，使用类似下面地址：

http://127.0.0.1:8080/get?imagepath=398351_1568345434461.jpg

本节介绍在 Spring Boot 2 中上传文件，以及在浏览器中显示图片的技术，主要思路是将上传文件代码服务化，以方便在多个地方复用，以及介绍浏览器显示图片的方式，各有利弊，可根据实际情况进行选择。以上代码请见本节源码。

13.2　Lombok 应用

在以面向对象方式的 Java 代码开发中，必不可少的是业务领域建模，正如前面章节中的多个示例，涉及不同的模型建立。但令开发人员不可避免和需要重复书写的是 Getter/Setter、ToString 等方法。尽管类似 Eclipse 开发工具提供了生成 Getter/Setter 方法的快捷菜单，一旦模型中字段增多，并且可能会有变动的情况下，后期再编辑是一件烦琐的事情。

本节介绍的 Lombok 是一款 Java 插件，可以让开发人员在开发中直接使用注解方式来代替一些冗余的代码，比如常见的 Getter/Setter 方法，能为开发人员节省构建时间。那么，对于这些省略的方法，Lombok 插件在编译源代码期间自动生成这些方法，不会降低代码运行的性能。

在 Spring Boot 2 项目中使用 Lombok 插件，需要手动下载该插件，下载地址如下：

https://projectlombok.org/download

将下载的 lombok.jar 放在 Eclipse 安装目录下，位置如图 13.5 所示。

图 13.5　lombok.jar 位置

打开命令提示符,进入 eclipse 所在位置,运行下面命令:

java -jar lombok.jar

运行命令后,打开如图 13.6 所示窗口。

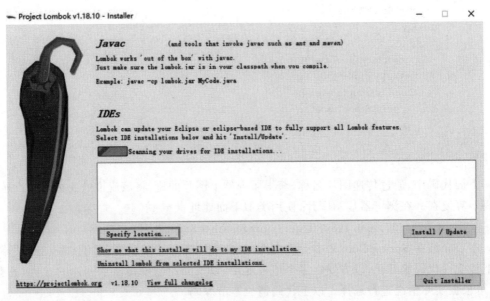

图 13.6 Lombok 界面

在图 13.6 中,单击 Specify Location 按钮,选择 Eclipse 所在当前目录后,单击 Install/Update 按钮安装即可。

查看安装是否成功,打开 Eclipse 配置文件 eclipse.ini,查看最后一行配置信息,如图 13.7 所示即可。

```
--add-modules=ALL-SYSTEM
-javaagent:D:\eclipse\lombok.jar
```

图 13.7 Eclipse 配置增加 lombok

以上操作只是为了防止错误发生,另一种安装方法是,直接编辑 eclipse.ini,在其最后一行加入 lombok.jar 所在位置即可。操作完成后,重启 Eclipse 使之生效。

在 Spring Boot 2 中使用 Lombok,需要引入相关依赖,即在 pom.xml 中加入下面依赖:

```xml
<dependency>
    <groupId>org.projectlombok</groupId>
    <artifactId>lombok</artifactId>
    <optional>true</optional>
</dependency>
```

之后就可以在项目中使用 Lombok 提供的注解。下面是一个简单示例代码:

```
@Slf4j
@SpringBootApplication
```

```java
public class C132Application {
    public static void main(String[] args) {
        log.info("启动");
        SpringApplication.run(C132Application.class, args);
        log.info("启动完成");
    }
}
```

上面代码中，在启动类名上，加入注解@Slf4j，那么这个类在编译的时候，变成如下：

```java
@SpringBootApplication
public class C132Application {
    private static final org.slf4j.Logger log =
        org.slf4j.LoggerFactory.getLogger(C132Application.class);
    public static void main(String[] args) {
        log.info("启动");
        SpringApplication.run(C132Application.class, args);
        log.info("启动完成");
    }
}
```

由上面可知，Lombok 是在源码编译期对代码进行了更改。而非 Spring 提供的那种注解，是在运行时用反射来实现业务逻辑。由此可知，为什么 Lombok 没有降低代码运行时性能。同时，利用好 Lombok，能减轻开发人员的烦琐工作。下面是 Lombok 提供的一些可用注解：

（1）@Setter/@Getter

以上注解可以应用在类名上也可以在属性上，将自动生成默认的 setter / getter 方法。下面是示例代码：

```java
@Getter
@Setter(AccessLevel.PROTECTED)
private String name;

@Setter
private int age = 12;

@Getter
@Setter
private String sex;
```

（2）@toString

一般，开发人员会手动拼接一个包含各字段的字符串用于调试。在类名上使用该注解，可以为该类自动生成一个包含各字段和值的字符串。

（3）@NonNull

以上注解用在一个字段上，用来判断是否为空，如果为 null，则抛出 NullPointerException 异常。

（4）@Data

以上注解应用在类名上，同时提供 getter、setter、equals、canEqual、hashCode、toString 等方法。

（5）@EqualsAndHashCode

以上注解应用在类名上，提供 equals、canEqual、hashCode 等方法。

（6）@Log

泛型注解，具体有很多种形式，比如前面介绍的@Slf4j。

（7）@AllArgsConstructor

（8）@NoArgsConstructor

（9）@Cleanup

（10）@RequiredArgsConstructor

（11）@Value

（12）@SneakyThrows

（13）@Synchronized

以上介绍了 Lombok 中提供的主要注解，利用好即能提升开发效率。但唯一缺陷，需要在 IDE 中安装一个 jar 插件，对于一些初次接触 Lombok 的开发人员来讲，不是太友好。但无论怎样，很多代码分享都使用了 Lombok 注解。基于此，开发人员应掌握如何在 IDE 中安装 Lombok、在 pom.xml 中如何引用该依赖的方法以及理解其常用注解。

13.3 热部署 Devtools 应用

为了进一步提高开发效率，特别是在开发中，开发人员需要多次重启服务以调试代码，对于这个烦琐工作，Spring Boot 提供了 spring-boot-devtools 模块，用来减少开发人员在开发过程中手动重启服务次数，加快开发进程。

spring-boot-devtools 模块支持 Java 文件修改后自动重启服务，实现类似 thymeleaf 模板实时编辑。在 Spring Boot 2 项目中，加入 spring-boot-devtools 模块的方法是，在 pom.xml 文件中，加入下面依赖：

```xml
<dependency>
    <groupId>org.springframework.boot</groupId>
    <artifactId>spring-boot-devtools</artifactId>
    <scope>runtime</scope>
</dependency>
```

保存配置文件 pom.xml 后，项目便具备热部署功能，并能在项目名称上看到变化，如图 13.8 所示。

由图所示，项目名称最后自动加上"[devtools]"标识，并在 Boot Dashboard 面板中有同样变化，如图 13.9 所示。

图 13.8　增加热部署项目　　　　图 13.9　Boot Dashboard 面板变化

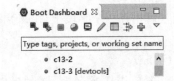

在图 13.9 中，项目名称后自动加上"[devtools]"标识。

通过以上操作，启动项目后，编辑或增加 Java 文件、编辑配置文件等，项目将在后台自动重启，并在 Console 面板中看到变化。

Devtools 支持以下目录的自动重新加载：

- /META-INF/maven；
- /META-INF/resources；
- /resources；
- /static；
- /public；
- /templates。

以上目录中文件发生变更后，因为不需要编译，将自动加载，而不是重启应用。如果要为其中某些文件夹禁用自动重新加载，则需要在配置文件 application.properties 中设置以下内容：

spring.devtools.restart.exclude = static/**

以上设置将排除/static，如果有多个文件夹需要排除时，则其间用逗号相隔。

如果有些文件不在类路径中，但又希望其改变后，同时能触发应用重新加载，则需要在配置文件 application.properties 中使用下面设置：

spring.devtools.restart.additional-paths = plugin/**

同样，使用下面设置项将设置一些文件夹内容被修改后，不触发自动重新加载应用。

spring.devtools.restart.additional-exclude = js/**

在开发现代前端先进技术中，已经支持代码变更，页面的自动刷新，具有很好的编程体验。那么，spring-boot-devtools 模块包含了内置 LiveReload 服务器，允许在资源变化后自动触发浏览器刷新。目前，LiveReload 浏览器扩展程序支持 Chrome、Firefox 和 Safari。下面介绍在 Firefox 中如何安装 LiveReload 浏览器扩展程序。

在 Firefox 中，单击附加组件项，将在页面中打开附加组件管理器页面，在其中搜索框中输入 LiveReload，如图 13.10 所示。

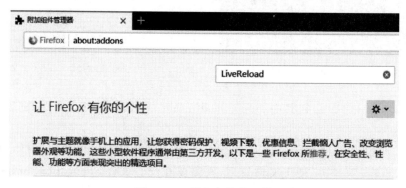

图 13.10 附加组件管理器

进行在线搜索,在查找结果中,找到 LiveReload 项,安装即可。安装完成后,将在浏览器的右边出现相关按钮,如图 13.11 所示。

图 13.11　LiveReload 按钮

该按钮默认是关闭的,当打开需要测试的页面后,单击该按钮,使其变绿,表示在工作中。此时,在 Eclipse 中,更改页面资源后,无须手动刷新,该页面会自动加载。节省开发人员不停刷新查看效果的时间,有效提升开发效率。

默认情况下,spring-boot-devtools 模块已启用实时重新加载 UI 页面,但如果想禁用该功能,则需要在配置文件 application.properties 中使用下面设置:

```
spring.devtools.livereload.enabled = false
```

13.4　更改应用启动 Logo

在启动 Spring Boot 2 应用时,会看到一个 Spring 的 Logo 标志,如图 13.12 所示。

图片比较经典,但开发人员一般希望使用自己个性图标,或是公司之类图标。那么,就需要替换掉 Spring Boot 2 启动默认 Logo。

其方法是,在项目的 resource 目录下,创建一个 banner.txt 文件,如图 13.13 所示。

图 13.12　Spring Boot 2 启用 Logo

图 13.13　banner.txt 文件位置

编辑 banner.txt 文件,在其中加入任何想要显示的内容,保存该文件,重启项目,即可在启动界面显示。

以上 banner.txt 文件放置为默认位置,如果希望将该文件放置到一个指定位置,比如一个自定义 banner 文件夹,便于文件管理,则需要在配置文件 application.properties 中设置 spring.banner.location 属性,示例代码如下所示:

```
spring.banner.location = classpath:/banner/mybanner.txt
```

通过以上配置可知，Logo 文件不但可以更改其所在文件夹，并且可以使用不同名称。

如果认为自己编辑一个好看 Logo 太复杂，有提供快速生成图标的页面，比如网站：

http://patorjk.com/software/taag/

以上是一个文字生成简单 Logo 的页面，如图 13.14 所示。

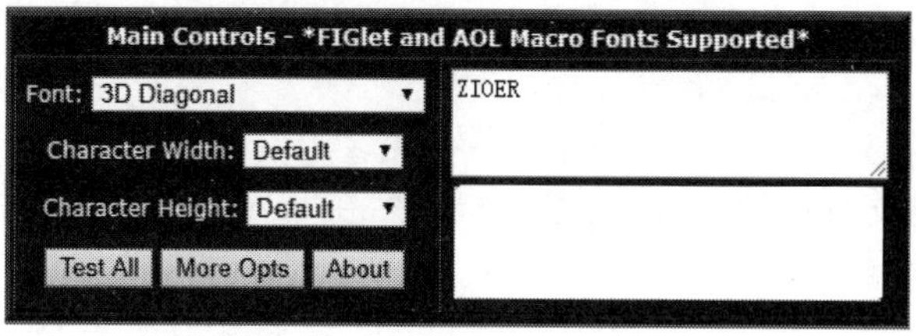

图 13.14　输入生成 Logo 的字体

在该页面输入文字，单击 Test All 按钮，即生成 Logo，然后选择需要的字体 Logo，将其复制到 banner.txt 文件中，重启应用后，Console 面板中显示如图 13.15 所示。

图 13.15　更改 Logo

除了更改 Logo 外，还可根据需要，更改 Logo 的颜色，方法是，在 banner.txt 文件中需要更改颜色的地方，放置下面属性配置：

${AnsiColor.BRIGHT_BLUE}

将上面属性配置放在 banner.txt 文件首行，则该 Logo 颜色全变为蓝色。支持更改的颜色为枚举，位于下面：

org.springframework.boot.ansi.AnsiColor

目前，支持的颜色包括如下：
- BLACK；
- BLUE；
- BRIGHT_BLACK；
- BRIGHT_BLUE；
- BRIGHT_CYAN；
- BRIGHT_GREEN；
- BRIGHT_MAGENTA；
- BRIGHT_RED；
- BRIGHT_WHITE；
- BRIGHT_YELLOW；

- CYAN；
- DEFAULT；
- GREEN；
- MAGENTA；
- RED；
- WHITE；
- YELLOW。

Spring Boot 2 还支持更换 Logo 背景，方法是，在 banner.txt 文件中需要更改背景颜色的地方，放置下面属性配置：

${AnsiBackground.BRIGHT_YELLOW}

通过以上方法，可以设置具有个性图标的启动 Logo。如果不想设置启动 Logo，则有三种方式进行设置，一是直接放置一个空的 banner.txt 文件在 resources 目录下。

二是在启动程序中编码设置，示例代码如下所示：

```
@SpringBootApplication
public class C134Application {
    public static void main(String[] args) {
        SpringApplication app = new SpringApplication(C134Application.class);
        app.setBannerMode(Banner.Mode.OFF);
        app.run(args);
    }
}
```

以上代码中，启动类 C134Application 中的 main 方法，调用 setBannerMode 方法，设置为 Banner.Mode.OFF。

三是在配置文件 application.properties 中设置，如下所示：

spring.main.banner-mode=off

13.5 应用缓存

在开发应用系统中，性能是很大一个问题，一个是网络带宽，另一个主要影响是应用系统的响应。一种有效的解决方案是使用缓存。

Spring Boot 2 提供了缓存机制，开发人员可以直接利用该缓存，另外也提供了外部缓存的方法，可以利用其他缓存方案进行缓存，加快应用的访问速度。本节将通过一个案例分析介绍 Spring Boot 2 的缓存。

建立一个 Spring Boot 2 工程，在 pom.xml 中加入以下依赖：

```xml
<dependencies>
    <dependency>
        <groupId>org.springframework.boot</groupId>
        <artifactId>spring-boot-starter-web</artifactId>
```

```xml
        </dependency>
        <dependency>
            <groupId>org.springframework.boot</groupId>
            <artifactId>spring-boot-starter-test</artifactId>
            <scope>test</scope>
        </dependency>
        <dependency>
            <groupId>org.springframework.boot</groupId>
            <artifactId>spring-boot-starter-data-jpa</artifactId>
        </dependency>
        <dependency>
            <groupId>org.projectlombok</groupId>
            <artifactId>lombok</artifactId>
        </dependency>
        <dependency>
            <groupId>com.h2database</groupId>
            <artifactId>h2</artifactId>
            <scope>runtime</scope>
        </dependency>
    </dependencies>
```

由以上代码可知，本示例使用了 JPA、H2 数据库以及 Lombok 等依赖，实际上，Spring Boot 2 相关依赖中，已经内置了 Hash Map 作为 Cache，如果没有关联其他第三方依赖时，可以直接使用其内置 Cache。图 13.16 是本项目结构示意图。

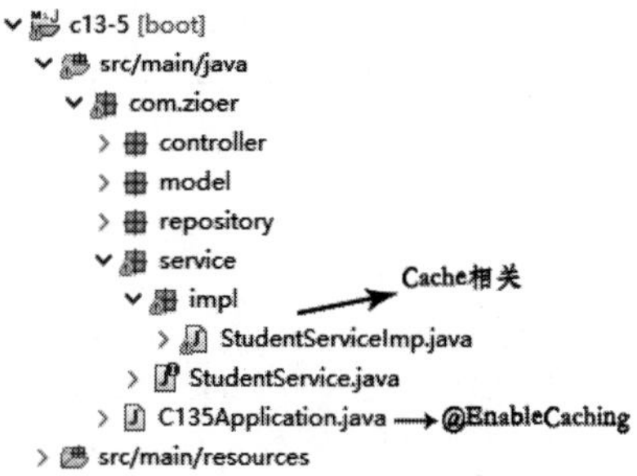

在图 13.16 中，标识了两个重点地方，即 Cache 启用和使用的地方，在项目的启动类文件 C135Application.java 中，加入了注解@EnableCaching，示例代码如下所示：

图 13.16　项目结构示意图

```java
@SpringBootApplication
@EnableCaching
public class C135Application {

    public static void main(String[] args) {
        SpringApplication.run(C135Application.class, args);
    }

}
```

注解@EnableCaching：启用缓存，如果不加入该注解，默认是不启用缓存。下面是服务层实现类 StudentServiceImp 中使用缓存的示例代码：

```java
@Service
public class StudentServiceImp implements StudentService {

    @Autowired
    StudentRepository repository;
```

```java
    @Override
    @CachePut(value = "student", key = "#student.id")
    public Student insert(Student student) {
        return repository.save(student);
    }

    @Override
    @CachePut(value = "student", key = "#id")
    public Student update(Student student) {
        return  repository.save(student);
    }

    @Override
    @CacheEvict(value = "student", key = "#id")
    public void delete(Student student) {
        repository.delete(student);
    }

    @Override
    @Cacheable(value = "student", key = "#id")
    public Student findById(int id) {
        Optional<Student> opt = repository.findById(id);
        return opt.isPresent()?opt.get():null ;
    }

    @Override
    @Cacheable(value = "student", key = "#name")
    public List<Student> findByName(String name) {
        return repository.findByName(name);
    }
}
```

在上面的代码中，应用到缓存中几个重要的注解，如下所示：

（1）@Cacheable

应用到方法上，即缓存该方法的结果。即首次查询时，缓存查询结果，以后再次访问时，将直接从缓存获取结果。该注解具有多个选项，如表13.1所示。

表 13.1 @Cacheable 相关选项

选 项	描 述
value	用于描述缓存的名称
key	缓存的关键字 key，可不指定，如果指定则要按 SpEL 表达式书写
condition	描述缓存的条件，可不指定，如果指定则要按 SpEL 书写，返回为 true 时进行缓存

在表13.1中，value 表示缓存名称，可不指定。该注解将在第一次查询时缓存，一般使用在查询方法上，比如方法 findById(int id)。下面是使用 condition 选项示例：

```
@Cacheable(value = " student", key = "#name", condition = "#name.length < 8")
public Book findByName(String name)
```

(2) @CachePut

该注解用于手动更新缓存,比如,对某一条记录进行更新后,同时更新对应缓存。这样,保证每次从缓存中读取数据总是更新后的内容。该注解具有多个选项,如表 13.1 所示。

提示:不建议将 @Cacheable 和 @CachePut 同时作用于同一个方法上,其实也没必要如此做法。

(3) @CacheEvict

该注解用于从缓存中清除指定内容,类似于 SQL 中的 delete 语句。该注解除了具有表 13.1 中的选项外,还具有表 13.2 中的选项。

表 13.2　@CacheEvict 独特选项

选　　项	描　　述
allEntries	设置是否删除所有缓存内容,默认为 false
beforeInvocation	设置是否在方法执行前删除所有缓存内容,默认为 false

下面是使用 allEntries 选项示例:

```
@CacheEvict(value = "student",allEntries = true)
public void deleteAll() {
    repository.deleteAll();
}
```

上面介绍的 3 个注解,基本能满足缓存的需要。保存上面示例项目,运行后,在 Postman 中测试,输入下面 URL:

http://127.0.0.1:8080/api/students/3

首次调用时,在 Console 面板中,可以直观看到查询语句,如图 13.17 所示。

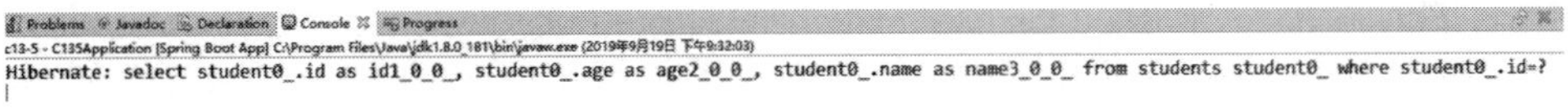

图 13.17　Console 面板 SQL 语句

接着,反复调用上面语句,在 Console 面板中,不再出现该 SQL 语句,表明是从缓存中读取数据。以此方法,可以测试更新缓存和删除缓存,查看效果,非常明显。

除了上面介绍几个注解外,还有几个有用注解,如下所示:

(1) @CacheConfig

该注解用在类名上,在上面几个示例中,都使用了 value 方法,并且相同,如此,可以将其定义在该注解中,减少重复书写,示例如下:

```
@CacheConfig("student")
public class StudentServiceImp implements StudentService {

    @Override
    @CachePut(key = "#student.id")
    public Student insert(Student student) {
        return repository.save(student);
    }
}
```

（2）@Caching

Cache 的组合使用，示例如下所示：

```
@Caching(put = {
    @CachePut(value = "student", key = "#student.id"),
    @CachePut(value = "student", key = "#student.name")
})
```

以及

```
@Caching(cacheable = {
    @Cacheable(value = "student", key = "#student.id"),
    @Cacheable(value = "student", key = "#student.name")
})
```

和

```
@Caching(evict = {
    @CacheEvict(value = "student", key = "#student.id"),
    @CacheEvict(value = "student", key = "#student.name")
})
```

以上三个示例代表三种使用组合方式。如果对某一个查询结果，不止对应一个 key 值时，此时，需要用到组合缓存方式。对于更新和删除情况，也用同样处理方法。

在本节示例中，缓存放置在服务实现层，而不是资源层，这是非常有好处的，一是屏蔽具体实现，即具体实现可以用关系数据库，或是使用非关系数据库；其次，对于调用层也无须了解缓存如何实现。

本节示例请查看完整源码，加深在 Spring Boot 2 项目中使用 Cache 使用方式和方法的印象。

本章小结

本章是很重要的一章，介绍在 Spring Boot 2 中几个在开发中使用到的知识点，实用性很强，并且每一节都有使用方法详细介绍、具体使用案例和源码。源码的目的，是加深对知识的理解与掌握，而不是简简单单运行一下源码看结果了事。正所谓，授人以鱼，不如授人以渔。

附录A Maven的使用

一些关于 Maven 的知识放在附录部分,主要是因为其与本书主题不太相关,但又是较重要知识,贯穿本书始终,故将其放置在附录中。

主流开发 Java IDE 一般都内置 Maven,比如 Eclipse 等,主要还在于其开源免费,谁都能获取。但是,Maven 同样在不断更新,很多开发人员还是愿意单独下载安装,然后配置。下面介绍在 Eclipse 中 Maven 的安装和基本命令。

A.1 Maven 安装

下载 Maven,建议在其官网下载最新稳定版本,地址如下:

http://maven.apache.org/download.cgi

下面以在 Windows 下安装进行说明。

在该页面,可以下载到最新版本的 Maven,建议下载已经编译后的二进制包,如图 A.1 所示。

Link	
Binary tar.gz archive	apache-maven-3.6.2-bin.tar.gz
Binary zip archive	apache-maven-3.6.2-bin.zip
Source tar.gz archive	apache-maven-3.6.2-src.tar.gz
Source zip archive	apache-maven-3.6.2-src.zip

图 A.1 下载链接

解压下载后的压缩包到本地计算机,比如 D:\programs\maven。
文件结构如图 A.2 所示。
至此,完成 Maven 的安装。在命令提示符下,进入其中的 bin 目录,输入下面命令:

mvn -version

运行结果如图 A.3 所示。

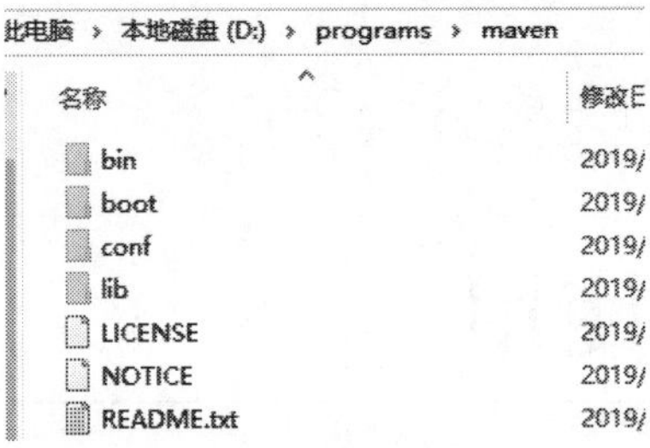

图 A.2　目录结构

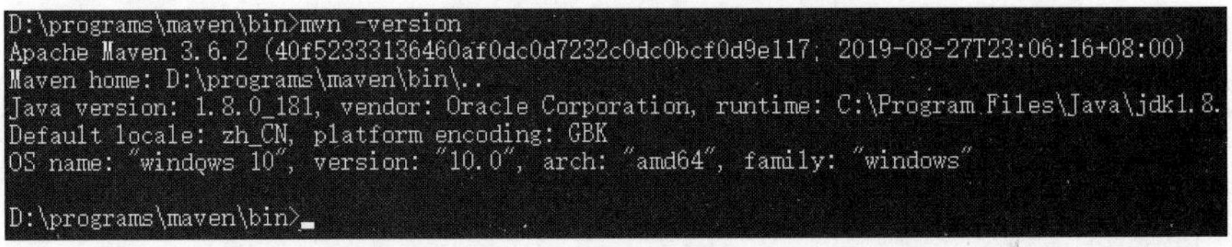

图 A.3　查看 Maven 版本

在图 A.3 中,可查看当前 Maven 的版本,并验证 Maven 安装成功。下面进行系统环境变量设置:在桌面的"我的电脑"上单击右键,选择"属性"命令,在打开的系统界面中,单击"高级系统设置"项,接着,在弹出的"系统属性"窗口中,单击"高级"选项卡,单击"环境变量"按钮,打开"环境变量"窗口,单击用户变量列表下的"新建"按钮,打开"新建用户变量"窗口,如图 A.4 所示。

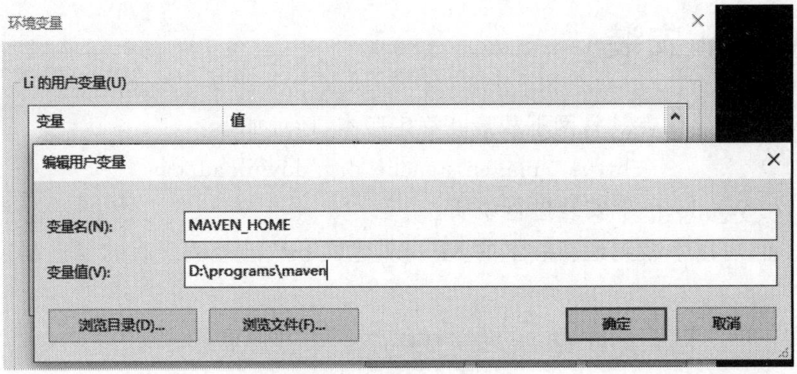

图 A.4　新建用户变量

在图 A.4 中,填写变量名和其对应的路径。编辑完成后,接着在"环境变量"窗口中,选择 Path 项,接着,单击"编辑"按钮,在"编辑环境变量"窗口中,增加 bin 路径,如图 A.5 所示。

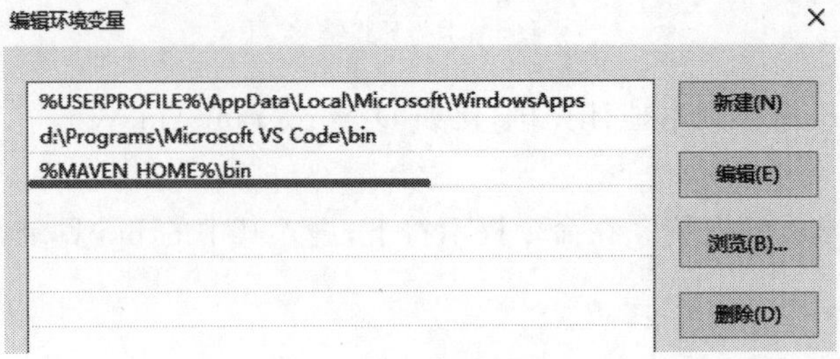

图 A.5　设置 Path

通过以上步骤,完成 Maven 的设置。至此,在命令提示符下,不必每次都必须进入 Maven 的 bin 目录中才能运行相关命令。

A.2 Maven 配置

Maven 的配置文件位于安装目录下的 conf\settings.xml,该配置文件是 xml 格式。下面是几个常用设置。

(1) 修改本地仓库所在位置

默认配置下,本地仓库位于当前用户目录下的.m2 文件夹中,可根据需要,更改默认的本地仓库位置。在 settings.xml 中,找到下面内容:

<localRepository>/path/to/local/repo</localRepository>

将以上内容复制到注释外,并更改其中的路径为本地仓库路径,示例如下所示:

<localRepository>d:/repository</localRepository>

以上设置将本地仓库路径更改为 d:/repository。

(2) 配置仓库镜像路径

默认情况下,中央仓库指向默认 http://repo1.maven.org/maven2/,但该仓库位于国外,在国内,访问速度很不稳定,或者,需要访问私有仓库,此时,需要更改仓库位置。在 settings.xml 中,找到 mirrors 元素,在其中加入下面内容:

```
<mirror>
    <id>alimaven</id>
    <name>aliyun maven</name>
    <url>http://maven.aliyun.com/nexus/content/groups/public/</url>
    <mirrorOf>central</mirrorOf>
</mirror>
```

以上设置,加入国内的某仓库,这样,所有的项目都默认到该仓库中寻找。同样,可以设置为私有仓库地址,适合于独立局域网中已有私有仓库。这种方法同时给所有项目访问仓库时,都更改了地址。

另一种方法是,给单独项目设置不同的仓库地址,此适合于部分特殊项目,其所需依赖位于不同的仓库中。设置方法是,在不同项目的 pom.xml 配置文件中,加入下面内容:

```
<repositories>
    <repository>
        <id>alimaven</id>
        <name>aliyun maven</name>
        <url>http://maven.aliyun.com/nexus/content/groups/public/</url>
    </repository>
</repositories>
```

以上方法,完成仓库地址的自定义。

A.3　Maven 基本命令

在使用 Maven 创建项目中，必须和 Maven 提供的命令打交道，尽管有时采用 IDE 界面能完成操作。

Maven 命令的基本格式如下所示：

mvn [option] ...

上面是 Maven 命令基本格式，即所有命令需以 mvn 开头，比如命令：

mvn - version

用于显示当前 Maven 版本。

下面是 Maven 提供的在项目中使用的命令：

（1）mvn compile

用于编译源代码，一般编译模块下的 src/main/java 目录。

（2）mvn package

用于项目打包，执行后，在项目的 target 目录中生成 jar 或 war 等文件。

（3）mvn clean

用于清理项目编译后的文件，即位于项目下 target 目录中内容。

（4）mvn test

用于测试命令，该命令将执行 src/test/java/下的 JUnit 测试用例。

（5）mvn deploy

该命令将打包的文件发布到远程仓库中，可以供其他开发人员使用。

（6）mvn install

该命令将打包的文件复制到本地仓库中，可供其他项目使用。

（7）mvn eclipse:eclipse

该命令将项目转化为 Eclipse 项目。

（8）mvn idea:idea

该命令将项目转化为 Idea 项目。

（9）mvn dependency:tree

该命令将打印出项目的整个依赖树。

（10）mvn site

该命令生成应用程序文档。

（11）mvn archetype:generate

用于创建 Maven 的普通 java 项目。

（12）mvn tomcat:run

用于在 tomcat 容器中运行 web 应用。

（13）mvn jetty:run

该命令调用 Jetty 插件的 run 命令，在 Jetty Servlet 容器中启动 Web 应用。

以上是 Maven 中基本使用命令,还可以带额外参数执行。

提示:在使用 Maven 命令时,需要在命令提示符下,进入项目 pom.xml 文件所在的目录,再执行。

下面是 Maven 中常用参数类型:

(1) -D[参数属性]

上面参数可通过-D 开头,给命令传入参数,比如:

```
mvn deploy -Dmaven.test.skip=true
```

以上命令在发布时,忽略单元测试。同理,下面命令是在打包时,忽略单元测试:

```
mvn package -Dmaven.test.skip=true
```

(2) -P[指定 profile]

如果项目中有多个 profile,在执行命令时,可以采用-P 开头,指定执行时的环境。比如:

```
mvn deploy -P dev
```

(3) -e

显示详细错误信息。

以上介绍的是 Maven 的基本命令。通过本附录,开发人员应该掌握 Maven 的基本操作,熟悉一些常用命令,以加快 Maven 项目的开发进程。

附录B YAML语法

在本书所有章节介绍中,Spring Boot 2 的配置文件为 application.properties,该文件的基本格式是"键=值",使用"#"开头作注释。Spring Boot 2 推荐使用 properties 文件来完成配置,同时,也推荐使用 YAML 文件来完成配置。

为了不让开发人员在阅读上产生困惑和思维上的转变,本书统一采用 properties 文件进行讲解,其好处是更直观、明了。但 YAML 文件也是一种常见的配置文件,并且采用工具很容易将 properties 文件转换为 YAML 文件,故在本附录中介绍 YAML 语法,并介绍转换工具命令的使用。

B.1 转换工具命令

在安装了插件 Spring Tool Suite 的 Eclipse 中,鼠标右键单击项目中配置文件 application.properties,在弹出菜单中,选择 Convert .properties to .yaml 命令,如图 B.1 所示。

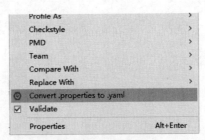

图 B.1 转换 YAML 命令

弹出转换对比窗口,如图 B.2 所示。

在该窗口中,可以较明显看到两个文件的对比,对于初学 YAML 语法者来说,这是一个很好的学习机会。接着,单击 OK 按钮,完成转换,并自动将文件名 application.properties 转换为 application.yml,其内容示例如图 B.3 所示。

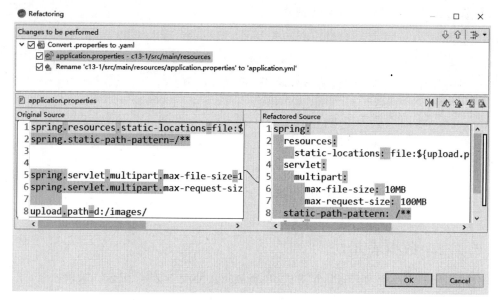

图 B.2　对比窗口

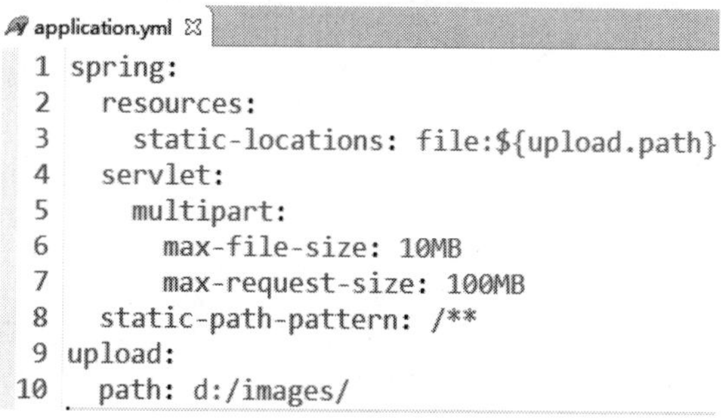

图 B.3　YAML 文件示例

下面介绍 YAML 语法,便于理解和掌握该语法。

B.2　基本语法

由图 B.3 所示,YAML 一个明显特征是缩进。语法规则如下:

(1) 使用缩进表示层级关系:缩进只能使用空格,不能使用 TAB,空格个数没有要求,但要求相同层级采用左对齐。

(2) 大小写敏感:例如 Spring 和 spring 代表不同。

(3) 使用#开头表示注释。

B.2.1　对象表示法

对象用冒号代表,格式为 key：value,同时,冒号后面有个空格。

对象的层级关系可以采用缩进表示：

```
upload:
  path: d:/images/
```

示例如图 B.3 所示，其更改为 properties 形式表示为：

```
upload.path=d:/images/
```

对象的另一种表示法：

```
upload: {path: d:/images/}
```

花括号内多个对象值之间用逗号分隔。

B.2.2 数组表示法

数组表示分为单行和多行显示方式，单行表示法使用中括号[]表示，示例如下：

```
key: [value1,value2,value3]
```

多行表示法使用减号-后跟空格和值形式来表示数组中的一个元素，同时注意每行缩进，示例如下：

```
key:
  - value1
  - value2
  - value3
```

以上简单介绍 YAML 几个重要的语法，重点在于抛砖引玉。由于有了转换工具，很容易理解和学习 properties 文件和 yaml 文件对应关系。

关于 YAML 更多内容，可关注网站 http://yaml.org。

附录C
IDEA工具介绍

IDEA 是 IntelliJ IDEA,是 Java 编程语言开发的 IDE。目前,其被大多数开发人员使用。IDEA 具有很多特点,包括:

(1) 深度智慧

IDEA 为源代码建立索引之后,会通过在各种情况下提供相关建议来提供快速、智能的体验:即时巧妙的代码完成、即时代码分析以及可靠的重构工具。

(2) 开箱即用的体验

诸如集成版本控制系统以及各种受支持的语言和框架之类的关键任务工具都应运而生,不包括插件麻烦。

(3) 智能选取

IDEA 提供基于语法的智能选择,默认快捷设置是 Ctrl+W,可以实现代码选取范围的不断扩充。

(4) 丰富的导航模式

IDEA 提供了丰富的导航查看模式。

(5) JUnit 的支持

IDEA 提供 JUnit 的支持。

(6) 历史记录功能

不用通过版本管理服务器,单纯的 IDEA 就可以查看任何工程中文件的历史记录,在版本恢复时可以很容易地将其恢复。

(7) 对重构的支持

IDEA 是所有 IDE 中最早支持重构的。

(8) 编码辅助

支持 Java 规范中提倡的 toString()、hashCode()、equals()以及所有的 get/set 方法,不用进行任何的输入就可以实现代码的自动生成。

(9) 灵活的排版功能

IDEA 支持排版模式的定制,可以根据不同的项目要求采用不同的排版方式。

(10) XML 的支持

XML 全提示支持。

(11) 动态语法检测

任何不符合 Java 规范、自己预定义的规范、累赘都将在页面中加亮显示。

(12) 代码检查

对代码进行自动分析，检测不符合规范的、存在风险的代码，并加亮显示。

(13) 完美的自动代码完成

智能检查类中的方法，当发现方法名只有一个时自动完成代码输入，从而减少剩下代码的编写工作。

以上仅列举部分 IDEA 工具部分特点。但需要注意的是，IDEA 提供了社区免费版和旗舰收费版，其中，社区版只提供基本功能，要想体验更好的 IDEA 工具，还是需要旗舰版，特别在面向 Web 开发，旗舰版集成了 Spring Initializr 等。

为了更好体验和使用 IDEA 工具，在官网上下载最新版本，其官网地址：

http://www.jetbrains.com/idea/

在下载页面中，提供了安装版和绿色非安装版两种方式，作者建议下载 Windows 非安装版。下面以使用旗舰版进行介绍。首次启动后，界面如图 C.1 所示。

图 C.1　IDEA 启动界面

单击 Create New Project 按钮，创建一个新工程，如图 C.2 所示。或在该界面中单击 Import Project 按钮，导入一个已存在的工程。

IDEA 旗舰版提供丰富的功能，从图 C.2 界面可看出，很多功能是社区版不具有的。单击 Spring Initializr 项，可以快速创建 Spring Boot 工程，图 C.3 所示为选择依赖界面。

按照操作，建立完成后的工程界面如图 C.4 所示。

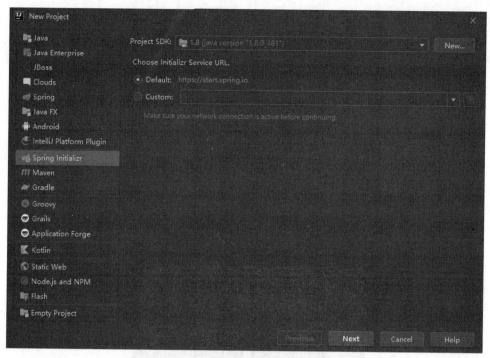

图 C.2　新建工程

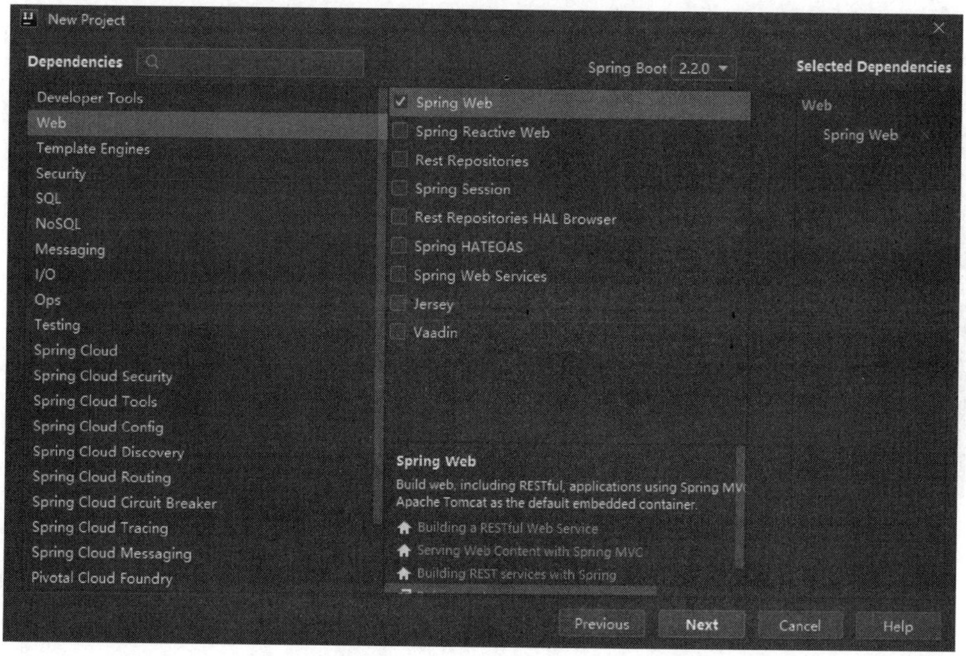

图 C.3　选择相关依赖

使用 IDEA 建立的工程目录和 Eclipse 建立的工程目录结构相同，只是在显示方式上有部分区别。等待 IDEA 下载完成相关依赖，然后，单击 IDE 界面右上角三角图标启动程序，如图 C.5 所示。

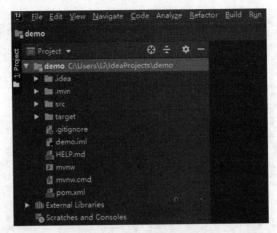

图 C.4　Spring Boot 2 工程

图 C.5　单击三角启动图标

如果单击小虫图标,表示调试工程。启动后,Console 界面中显示熟悉的 Spring Boot Logo,如图 C.6 所示。

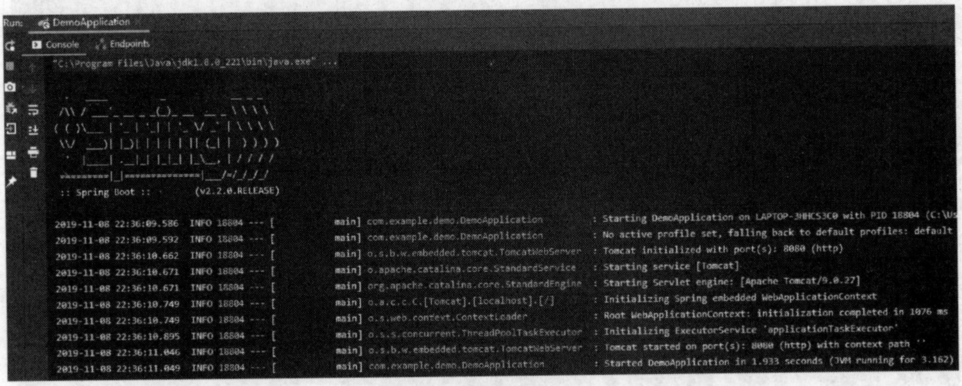

图 C.6　Console 显示

至此,完成了 IDEA 中 Spring Boot 2 项目从创建到运行的介绍。IDEA 提供的更多功能需要开发人员在实际操作中进行体验。

附录D Tomcat服务器

在部署 Java Web 应用时，Tomcat 服务器作为开源项目，很受开发人员喜爱，并被广泛使用。Tomcat 能被在多种操作系统中安装，下面以在 Windows 下安装为例介绍。

下载 Tomcat 最新稳定版，下载地址如下所示：

https://tomcat.apache.org/download-90.cgi

下载 Windows 对应版本，比如 64 位 Windows，下载类似下面版本：

apache-tomcat-9.0.34-windows-x64.zip

将下载的文件解压到本地目录，如图 D.1 所示。

图 D.1 Tomcat 解压本地

下面，配置 Tomcat 环境变量，打开本地环境变量配置界面，位于"计算机→属性→高级系统设置→高级→环境变量"。

在用户变量中添加以下变量和内容：

TOMCAT_HOME: C:\programs\tomcat\apache-tomcat-9.0.34

示例如图 D.2 所示。

接着，新建变量 CATALINA_HOME，值与 TOMCAT_HOME 的值一样，如图 D.3 所示。

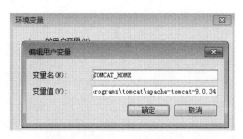

图 D.2　TOMCAT_HOME 变量设置

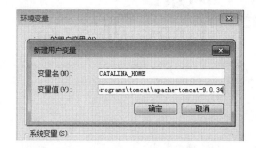

图 D.3　CATALINA_HOME 变量设置

下面，修改用户变量 Path，添加如下内容：

;%TOMCAT_HOME%\bin;%CATALINA_HOME%\lib

提示：Path 中各个值之间一定要用分号";"分隔。

至此，Tomcat 服务器安装完成，在命令提示符下输入下面命令：

startup

出现如图 D.4 所示界面，启动 Tomcat 服务成功。

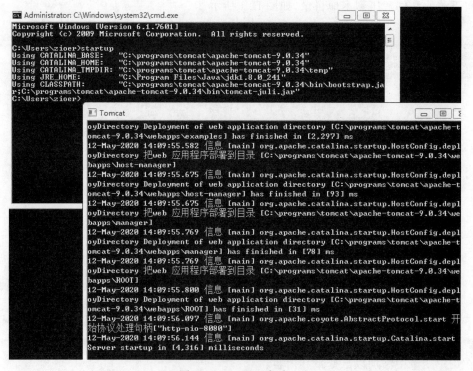

图 D.4　Tomcat 启动

在浏览器中浏览下面地址：

$$\text{http://127.0.0.1:8080}$$

将出现如图 D.5 所示界面。

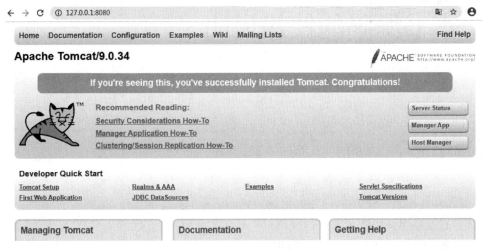

图 D.5　Tomcat 界面

提示：Tomcat 9 需要本地安装 Java 8，否则启动失败。

附录E
本书源码的使用说明

随着各种新技术涌现，Jar 包管理更加成熟，应用范围越来越广，作者本人也热衷于代码开发，一直在使用新的技术来进行代码重构和编写。并且，互联网发展迅速，网速也比以前快了很多，Java 程序所需依赖下载速度也快了很多。鉴于此，本书分享的源码所需依赖 Jar 包全部采用 Maven 方式管理。

其好处是源码体积变得很小，分享更加容易。但开发人员同时需要掌握 Maven 工具的使用，好在本书附录中，提供了 Maven 使用简单说明，掌握其使用语法也比较容易。

本书提供所有章节中示例源码，开发环境基于开源、Java 的可扩展开发平台 Eclipse，但源码也可在 MyEclipse、IDEA 等工具中使用。下面以 Eclipse 中如何使用源码进行说明。

单击 Eclipse 中 File 菜单，选择 Import…命令，打开 Import 菜单，如图 E.1 所示。

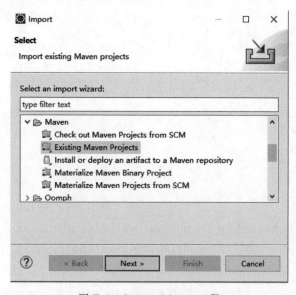

图 E.1　Import Maven 工程

选择列表中 Maven→Existing Maven Projects 后，单击 Next 按钮，打开选择待导入源码文件夹窗口，如图 E.2 所示。

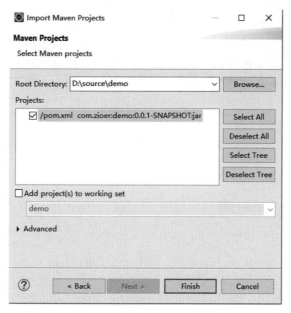

图 E.2　选择需要导入的源码

在该窗口中，在 Root Directory 输入框中输入源码中项目的根目录，同时勾选该 pom.xml 文件，单击 Finish 按钮，完成项目的导入。导入过程中，需要等待工程下载相关依赖，下载完成后，源码导入结束，在 Project Explorer 面板中可看到刚导入的源码工程，如图 E.3 所示。

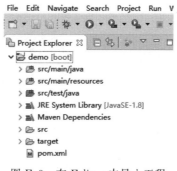

图 E.3　在 Eclipse 中导入工程

提示：导入源码过程，需要在连接互联网情况下进行，否则可能会导入源码失败。在 MyEclipse 中导入源码过程和上面类似。